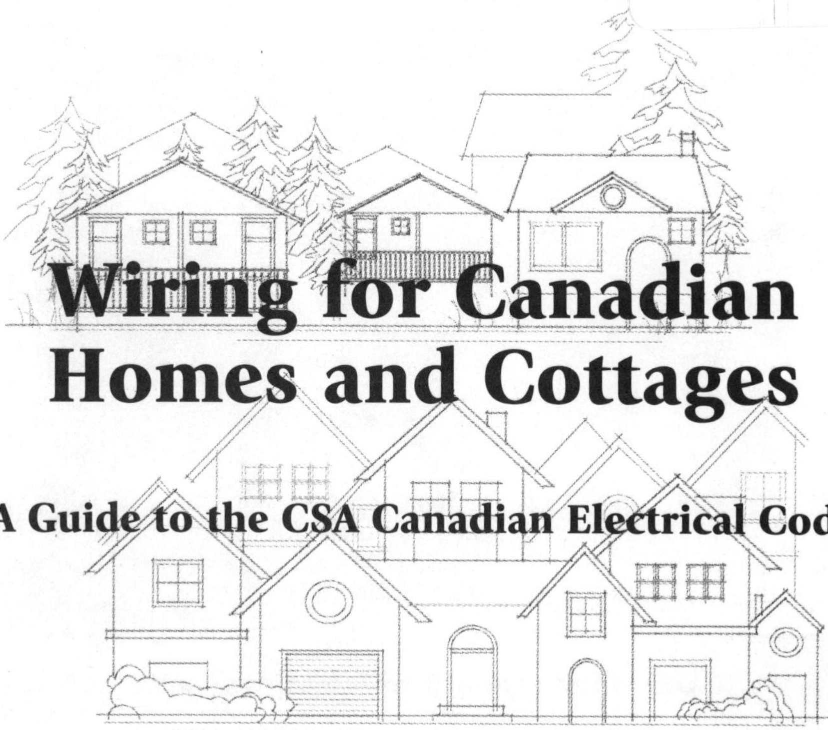

# Wiring for Canadian Homes and Cottages

## A Guide to the CSA Canadian Electrical Code

CANADIAN STANDARDS ASSOCIATION

Published by Canadian Standards Association
A not-for-profit private sector organization
5060 Spectrum Way, Suite 100, Mississauga, Ontario, Canada L4W 5N6
1-800-463-6727 • 416-747-4044



CANADIAN STANDARDS ASSOCIATION



ISBN 1-55397-522-7

# CONTENTS

# USING THIS BOOK

This book is intended to help the reader understand the residential occupancy requirements found in the 2002 *Canadian Electrical Code, Part I.* It is important to emphasize that this book is not intended to be used as a manual for electricians, renovators, contractors, or do-it-yourselfers on the installation of the electrical wiring systems required in a home or cottage. Most household electrical projects require the tools, skills, and experience of certified electricians. The Code also requires that when an electrical installation is installed or altered, the municipal, provincial, or territorial inspection department that has jurisdiction must inspect it to determine whether it complies with the Code and with any local requirements. This book must not be taken as a substitute for the *Canadian Electrical Code,* municipal, provincial, or territorial requirements, or the expertise of licensed electrical contractors, certified electricians, and inspection authorities.

# INTRODUCTION

## Up to Code

Since the 1890s, Canadians have enjoyed the benefits of residential electrical supply. At first electricity was used in the home principally for lighting, but by the 1920s the rapid development of a wide range of electrical appliances led to increased demand and increasingly complex home wiring requirements. By the middle of the twentieth century, the growth of large-scale generation and transmission facilities, vigorous provincial electrification programs, and a post-war acceleration in the industrial production of washing machines, power tools, electric ranges, refrigerators, and other kitchen appliances combined to create the consumer electricity market we take for granted today. And in the last fifty years, the range of residential and consumer needs has continued to grow. Today, residential consumers use electricity for far more than lighting and appliances — they may have the power needs of a spa (hot tubs, saunas, swimming pools, whirlpool baths), a restaurant kitchen (high-end ranges, heavy-duty refrigerators and freezers), an office (computers and networks, printers, copiers, fax machines), or a private theatre (sound systems, media rooms, home theatres).

Today, the *Canadian Electrical Code* (referred to throughout this Guide as "the Code"), now in its nineteenth edition, with more than 500 pages and indexed topics ranging from arc fault circuit interrupters to yacht clubs, represents more than 75 years of continually updated and improved consensus among regulatory authorities, insurers, inspection authorities, electricians, and electrical manufacturers on the safe installation, operation, and maintenance of electrical equipment. Each new edition has benefited from new knowledge of developing technology, the lessons of experience, and the best practices of experts in the field. As a result, in Canada, when a job is "up to Code," it satisfies some of the most stringent and carefully developed requirements in the world.

# Terminology

The *Canadian Electrical Code* defines many technical terms and uses others that are common in related trades and industries. This book introduces and illustrates many of these important terms as they apply to homes and cottages. A glossary has been provided at the back of the book for handy reference.

One set of terms is especially important in the discussion of residential electrical service. The Code's requirements are often based on a distinction between different kinds of residence: a "residential occupancy," a "dwelling unit," and a "single-dwelling unit."

## Residential occupancy

"Residential occupancy" refers to the occupancy or use of a building, or a part of a building, by a person who comes and goes at will and for whom sleeping accommodation is provided. Examples include rooms in a hotel, in a dormitory, or in a retirement home, so long as these rooms do not have cooking and eating facilities. (A residential occupancy might have a mini-fridge, a microwave oven, and/or a small eating area, but it would not have what would be described as a kitchen — a room or area used primarily to prepare and cook food.) The term does not apply to accommodations in hospitals or correctional facilities.

## Dwelling unit

A "dwelling unit" refers to one or more rooms that are used by a person as a housekeeping unit, with cooking, eating, living, and sleeping facilities. A dwelling unit contains a kitchen, as well as sleeping and living areas or rooms. Examples include apartment building suites, flats, and lofts.

## Single-dwelling unit

A "single-dwelling unit" is a dwelling unit that has access to outdoor areas at ground level and that may or may not have an attached garage or carport. A single-dwelling unit may be a detached house, a unit of a semi-detached house, a unit of row housing, or a unit of a duplex, triplex, or quadruplex house.

This book is about homes and cottages that fall under the definitions "dwelling unit" and "single-dwelling unit." The general term "dwelling" is used throughout the book, except when it is important to make a distinction between these types of residential settings.

## Bringing the Code home

There are a lot of experts at work on electrical safety in Canada. But making electricity safer — especially in the home — is not something you can simply leave to the experts. Every homeowner, consumer, tenant, and landlord should have some familiarity with residential electrical installations and should be aware of the basic requirements for receptacles, lighting, large appliances, carbon monoxide detectors, smoke alarms, and other equipment.

A first-time reader might find the *Canadian Electrical Code* a bit challenging; it is, after all, a highly technical and detailed document that covers almost every imaginable aspect of electrical installations, not only in residences but in factories, industrial settings, airports, marinas, and a host of other environments. But electrical requirements are not something that only the experts can understand. That is why we have produced this book — an accessible survey for non-specialists of the Code requirements for residential electricity. It is designed to

- help prospective home and cottage buyers design home electrical systems to meet the Code requirements and to anticipate demands that will be placed on the home's electrical installation in the future;
- help prospective home and cottage buyers assess the condition of wiring and electrical equipment in new and older homes;
- help home and cottage owners spot potential problems, plan effective renovations, and deal more productively with electricians and service providers;
- help landlords ensure that they provide adequate electrical service in rental properties; and
- help tenants understand what can — and cannot — be done with existing electrical equipment and wiring.

In all these areas, *Wiring for Canadian Homes and Cottages* is intended to describe what has to be done — and how it has to be done — to be "up to Code."

The chapters of this book follow electricity as it enters a dwelling through the service connection, travels to the customer's service box, is divided into branch circuits in the panel, and then is distributed throughout a dwelling through wiring for receptacles, lighting, and smoke alarms. The survey of residential wiring includes special requirements for hallways, rooms, finished walls, walkways, porches, etc., because the safety of an electrical installation often depends on such factors as accessibility, exposure to weather or accidental contact, and construction methods. The book also deals with major electrical equipment and special electrical installations, reviewing the Code requirements for kitchen equipment, heating and air-conditioning equipment, and spas, pools, saunas, and hot tubs. At the end of the book, a glossary and a number of worksheets have been provided, as well as "at a glance" summaries of the Code requirements for receptacles, lighting, smoke alarms, carbon monoxide detectors, pools, tubs, and spas.

This book surveys Rules for residential electrical installations found in Sections 2, 4, 6, 10, 12, 14, 16, 26, 28, 30, 32, and 62 of the *Canadian Electrical Code, Part I* (nineteenth edition, 2002). The Rules of the Code are expressed in metric measurements; imperial measurements have been added in parentheses in this book as a convenience for the reader.

## About CSA

CSA is a membership-based association serving industry, government, consumers, and other interested parties in Canada and the global marketplace. A leading developer of standards and codes, CSA aims to enhance public safety, improve quality of life, preserve the environment, and facilitate trade. To help people understand and apply standards, CSA offers information products and training. The Canadian Standards Association is a division of CSA Group, which also includes CSA International for product testing and certification and QMI for management systems registration.

# 1

# SERVICE INSTALLATIONS

## What are the basic requirements for service installations?

### How is residential electricity supplied?

Electrical service is supplied by the "supply authority" — the term the *Canadian Electrical Code* uses to refer to any person, firm, corporation, company, commission, or other organization that supplies electrical energy by a set of conductors running from the supply authority's mains to a consumer's service. These conductors — the "supply service" — belong to the supply authority, and the supply authority is responsible for servicing and maintaining these conductors after initial installation. The "consumer's service" refers to the portion of the consumer's installation between the supply authority connection and the service disconnect inside the building where the consumer's service conductors terminate. The consumer's service, which includes conductors, metering equipment, and the service disconnect, brings electrical energy into the dwelling. It can then be distributed through a panel to the branch circuits to meet the consumer's individual electrical requirements.

To prevent unintentional or accidental contact, the supply service is located either overhead or underground (see Figure 1). An overhead supply service is more easily accessible, and more easily provided and modified. In many cases, the supply authority installs its equipment at no extra charge to the consumer. In the case of an underground supply service installed as part of a new construction, however, the consumer often pays for the materials and installation as part of the overall building costs. The supply authority is still responsible for the service and maintenance of the underground service supply conductors. It is always important to determine — and comply with — the supply authority's requirements for a service installation, but this is especially true in the case of underground installations, where repairs and changes can be very costly and inconvenient.

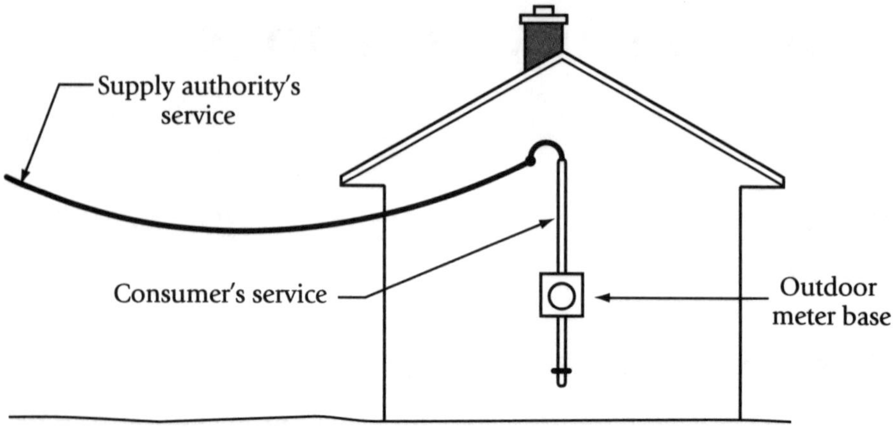


**Overhead service installation**

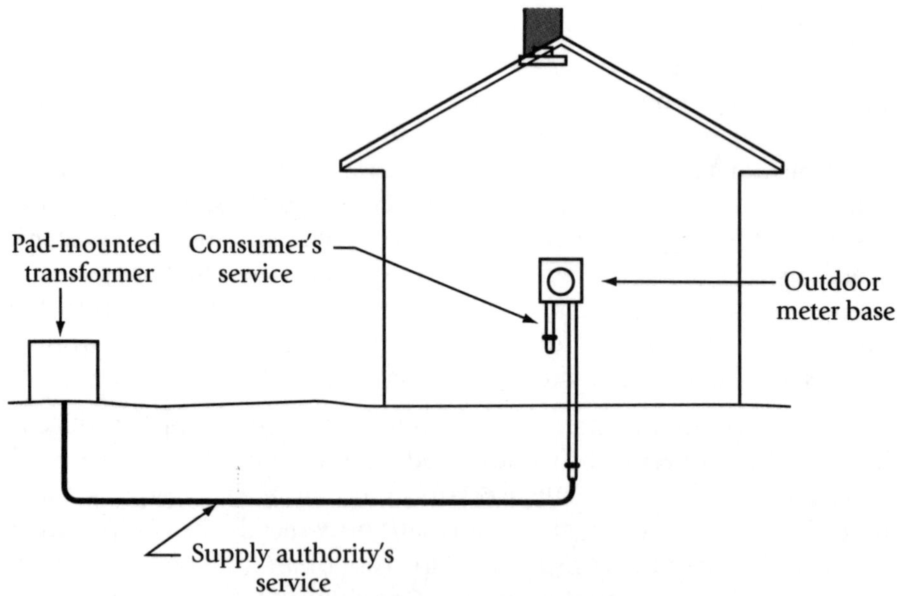


**Underground service installation**

Figure 1
**Supply service**

When preparing for a new installation, it is essential to work with the supply authority and have a well-developed plan. The following checklist summarizes the steps in the process and the key information that will be required.

**Step 1** Determine the size of service the dwelling will require (see "How is the size of service calculated for a single dwelling?" on page 40).

**Step 2** Arrange a meeting with a representative from the customer service or "service locate" department of the supply authority that will provide the electrical energy to the dwelling. For this meeting, assemble

- information and calculations related to the planned service size;
- a site plan of the dwelling;
- a set of architectural drawings for the dwelling; and
- any other information the supply authority requests.

**Step 3** In consultation with the supply authority's representative, determine the following:

- the location of the service meter at the dwelling;
- the type of meter base into which the supply authority will install its meter;
- the size of meter base, in amperes;
- the mounting height of the meter base;
- for an overhead service:
    - the location of the point of connection (rack);
    - the type of insulated rack;
    - the required number of insulators;
    - the mounting height of the rack;
    - the routing of overhead supply service conductors; and

  - any costs that the customer must pay [these costs are usually incurred when the length of overhead conductors on private property exceeds 30 m (100 ft)];
- for an underground service:
  - the routing of the underground trench from the meter on the dwelling to the connection point of the supply authority's underground conductors (in cottage areas, supply authorities may require armoured cable to be used where there is exposed rock and little soil);
  - the depth of the trench;
  - the method of installing underground conductors — direct burial or raceway (sealed channels). If direct burial, the type of conductors or cable that must be used;
  - for an underground raceway:
    - the type of raceway;
    - who will supply the underground conductors;
    - who will install the underground conductors; and
    - whether a wire or rope will be required for the supply authority's installation of the conductors;
  - the length of underground conductor to be left for termination by the supply authority; and
  - any costs to the customer.

## What are the basic location requirements for a supply service?

Accessibility and safe working conditions are the basic requirements for locating a service installation. The supply authority is responsible for installing, connecting, maintaining, servicing, and removing the service supply conductors. As a result, the supply authority will choose a location that helps ensure safe and adequate working space for its personnel, as well as for other workers who may service, operate, or maintain the consumer's service equipment.

In selecting a location, the supply authority must also consider specific requirements set out in the Code. The location must be clean and environmentally appropriate for the electrical equipment that is to be installed. The latter requirement refers to the Code's classification of environments as "dry," "damp," "ordinary," "outdoor," or "wet", and the corresponding designations of enclosure types. Enclosures are boxes used for electrical devices, outlets, and connections. Table 1 summarizes the Code's enclosure classifications.

| Location type | Description of location | Enclosure designation |
|---|---|---|
| Dry | Not normally subject to dampness | Type 1 |
| Damp | Normally or periodically subject to condensation or moisture in, on, or adjacent to electrical equipment | Type 1 |
| Ordinary | At normal atmospheric pressure and under normal conditions of use, electrical equipment is not exposed to damage from mechanical causes, excessive dust, moisture, or extreme temperatures, and electrical equipment is entirely free from the possibility of damage through corrosive, flammable, or explosive atmospheres | Type 1 |
| Outdoor | Exposed to weather | Type 3 |
| Wet | Liquids might drip, splash, or flow on or against electrical equipment | Type 2 — drops of falling liquid due to condensation or other causes<br><br>Type 4 — subject to direct streams of water |

**Table 1**
**Enclosure types**

In most dwellings, the service equipment is installed indoors in an "ordinary" location, and a Type 1 (or "general-purpose") enclosure is used. The Code does not require such an enclosure to be labelled with its Type designation.

The accessibility of the service installation must be balanced with the need to limit exposure to live electrical energy. The consumer's service conductors are usually not protected by an overcurrent device until they exit the service disconnect. The supply and consumer's service conductors, although insulated, carry electricity at a level that could pose a potential fire or shock hazard to a dwelling or its occupants. The Code requires that the distance from the conductors' entry point into the dwelling to the service disconnect be kept as short as possible. In addition, where the conductors run along or through walls and other surfaces, they must be sheathed or enclosed in raceways to prevent direct contact. The hole in the outer or exterior building wall, through which the consumer's service raceway or cable enters the building, must be sealed to prevent the entrance of moisture, insects, etc. See Figure 2.

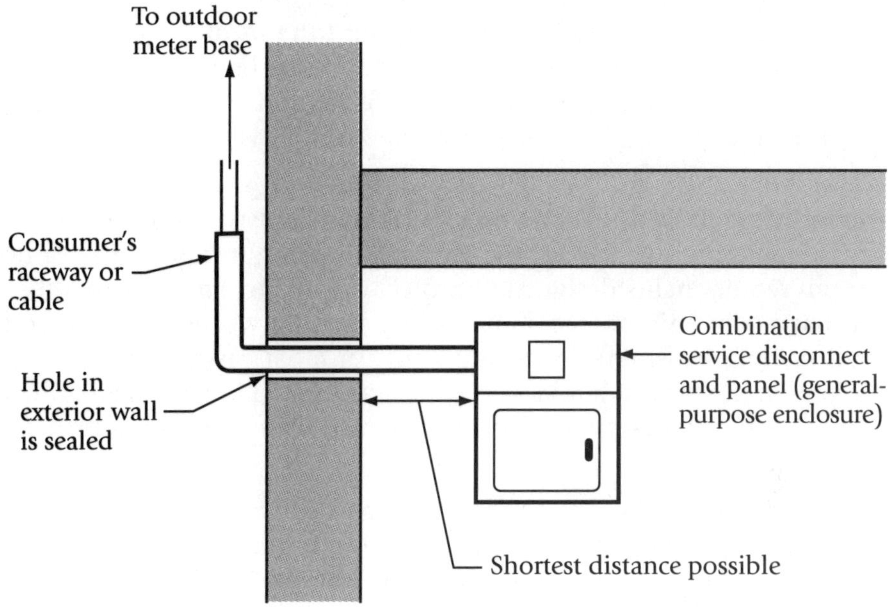


**Instead of a combination service disconnect and panel, a separate service disconnect and panel may be used, as shown below:**

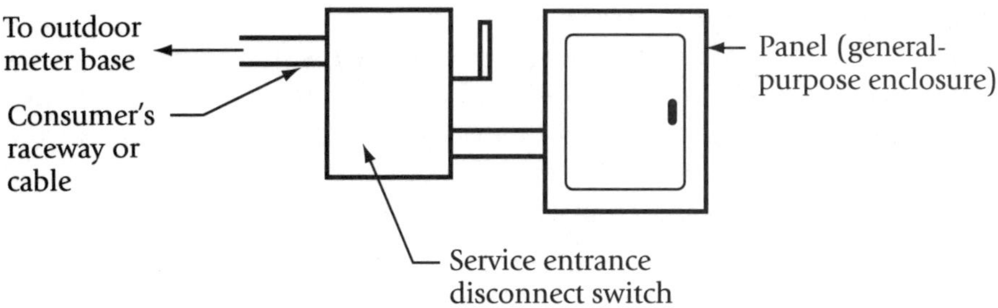


Figure 2
**Consumer's service conductors**

When locating and installing the service installation for a dwelling, the supply authority will also take into account the location of metering equipment. The meter records, in kilowatt-hours, how much electrical energy is used in the dwelling, and this information is used when billing the consumer for the cost of supplying and delivering the electrical energy.

The location chosen for metering equipment must also be clean and environmentally suitable for the equipment being installed. The Code sets out requirements designed to protect workers by ensuring that they will have convenient access and secure footing while servicing and reading electrical metering equipment. In most cases, the metering equipment in new residential occupancies is located outside. This allows supply authority personnel to service and read the meter without having to enter the dwelling.

Wherever practicable, the service raceway or cable should enter the building above the ground. A raceway that enters a dwelling below ground level can allow the entrance of moisture and underground gases, which pose health risks, create fire hazards, and can cause a rapid deterioration of electrical service equipment.

Whether the raceway is above or below ground level, the Code requires that it be drained and sealed and that precautions be taken to prevent the entrance of moisture and gases into the dwelling. A number of products on the market, such as insulating putty, can be used to prevent the movement of air or gases through the raceway. The local electrical inspection department or an electrical supplier can recommend a product that will not cause deterioration of the insulation on the service conductors. The sealant can be applied on either the exterior or interior side. Draining can be provided by drilling a 6 mm (¼ in) hole in the bottom of the LB fitting outside the exterior wall and sloping the raceway slightly upwards to meet the interior service equipment (see Figure 3).

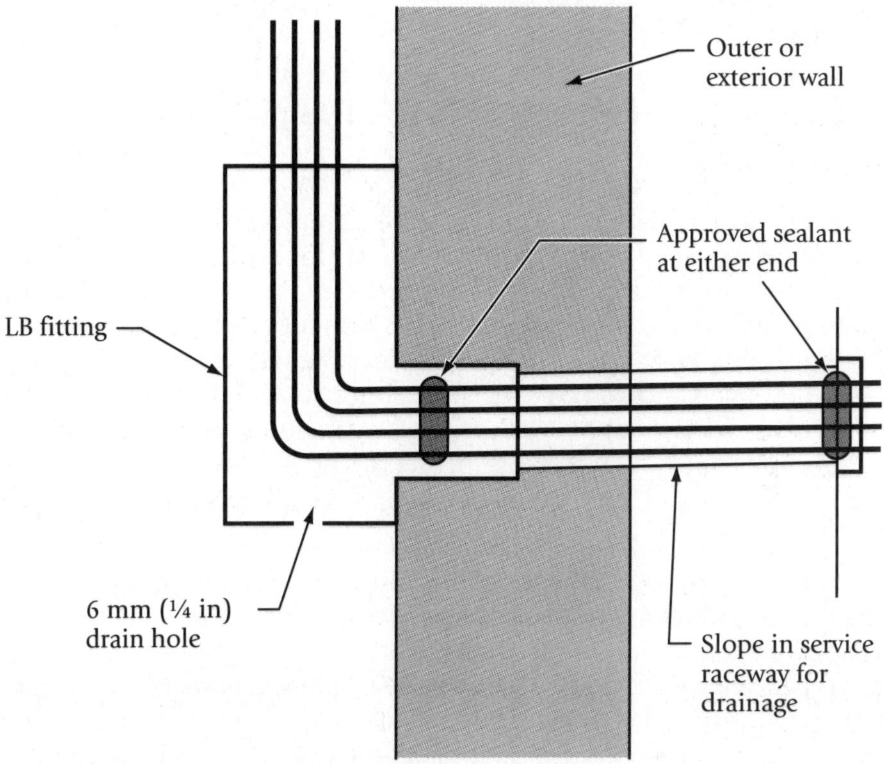


Figure 3
**Service conductors passing through exterior wall**

# What are the requirements for overhead supply services?

## What are the requirements for the point of attachment?

An overhead service installation puts the supply service conductors out of people's way by installing them a minimum distance above the finished ground level. To keep them there, the Code requires that the point of attachment of the supply service conductors to the dwelling be mechanically secure. These requirements take into account the possibility that the combined weight of the conductors and any snow, ice, or wind loads on the supply authority's or consumer's overhead service conductors could cause the conductors to sag, pull the point of attachment (the insulated rack) out of the wall, or bend the service mast.

The intent of the Code requirements is to allow access for service and maintenance and to prevent people and vehicles from contacting the overhead conductors. To allow the servicing of overhead supply service conductors, the point of attachment to the dwelling must not be more than 9 m (30 ft) above ground level. Minimum heights above ground level are also specified to help prevent contact with people or vehicles on the ground. The minimum clearance between the supply service conductors and the ground is determined by what might pass beneath the conductors. The Code requirements for the minimum ground clearance are

- 5.5 m (18 ft) when run across highways, streets, lanes, and alleys;
- 4 m (12 ft) when run across driveways to residential garages; and
- 3.5 m (11 ft) when run across ground normally accessible to pedestrians only.

The point of attachment for the supply service conductors on the dwelling is always higher than the values listed above because the minimum clearance is measured from the lowest point of the run of supply service conductors (see Figure 4). A practical rule in the industry is to install the point of attachment for service conductors at least 300 mm (1 ft) higher than the applicable minimum ground clearance.

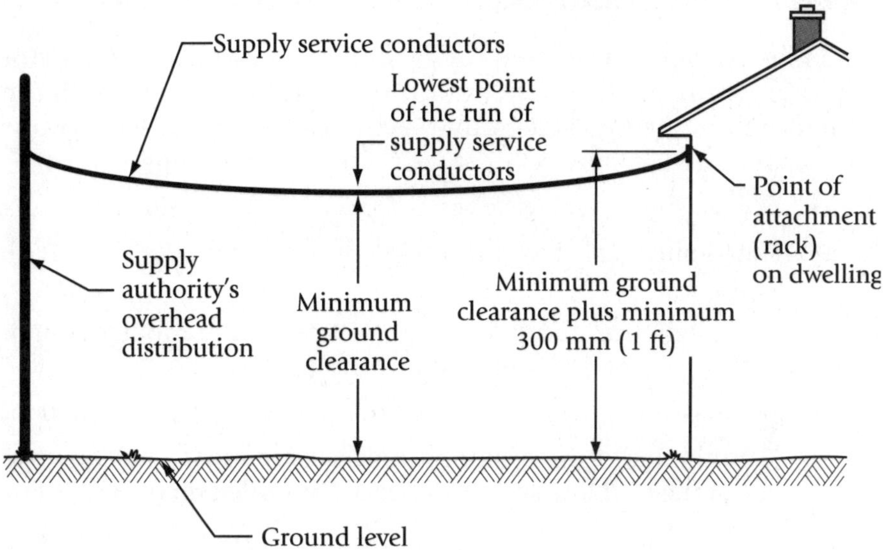


**Figure 4**
**Minimum clearance for overhead conductors**

The bolts securing the rack must not be smaller than 9.5 mm (3⁄8 in). Depending on the size and length of the supply service conductors, larger bolts may be required. If the point of attachment is being secured to wood to support the structure, the wood must not be less than 38 mm (1½ in) in any dimension. The local supply authority may also have its own minimum requirements for securing the rack.

To prevent people from contacting exposed supply service conductors that may be outside the cable sheath or raceway at windows, doors, and porches, a minimum clearance of 1 m (3 ft) is required.

## What are the requirements for a service mast?

When the point of attachment for overhead supply service conductors cannot be made on the exterior of the dwelling because of insufficient clearance from the finished ground level, a service mast (see Figures 5 and 6) may be used, provided that the following conditions are met:

- The consumer's service does not exceed 200 A or 750 V.
- The supply service conductors do not exceed 30 m (100 ft) in length.
- The maximum unsupported projection of the tubular support above the roof does not exceed 1.5 m (5 ft).
- The maximum load on the mast does not exceed 270 kg (600 lb).
- A minimum of three support brackets are used, with the upper one at the roof line and the other two installed on the wall of the building.
- There is a minimum clearance of 915 mm (3 ft) between the roof and the supply service attachment (the rack). This dimension may be reduced to 600 mm (2 ft) between the bottom of the drip loops and the roof (see "What are the requirements for a service head?" on page 15). This reduction to the 915 mm distance is not allowed in areas where there is heavy snowfall.

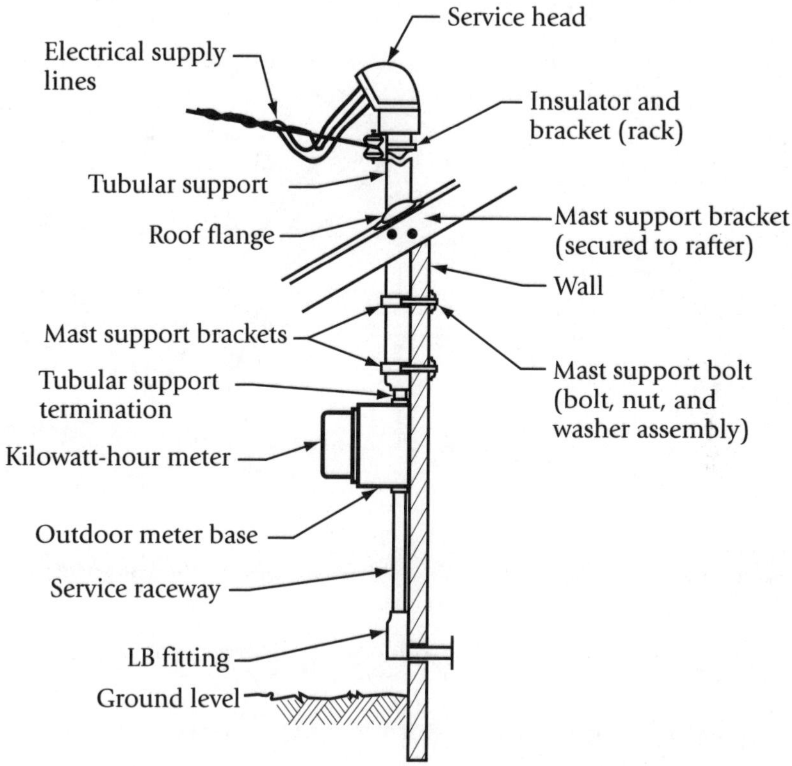


**Figure 5**
**Service mast**

The components of an approved service mast include

- tubular support (the tubular support is made of high-strength steel; rigid conduit is not acceptable);
- support brackets (standard conduit support straps are not strong enough);
- roof flange (to keep water from running down the tubular support into the roof space);
- supply service attachments (racks);
- service head; and
- tubular support termination (at the lower end).

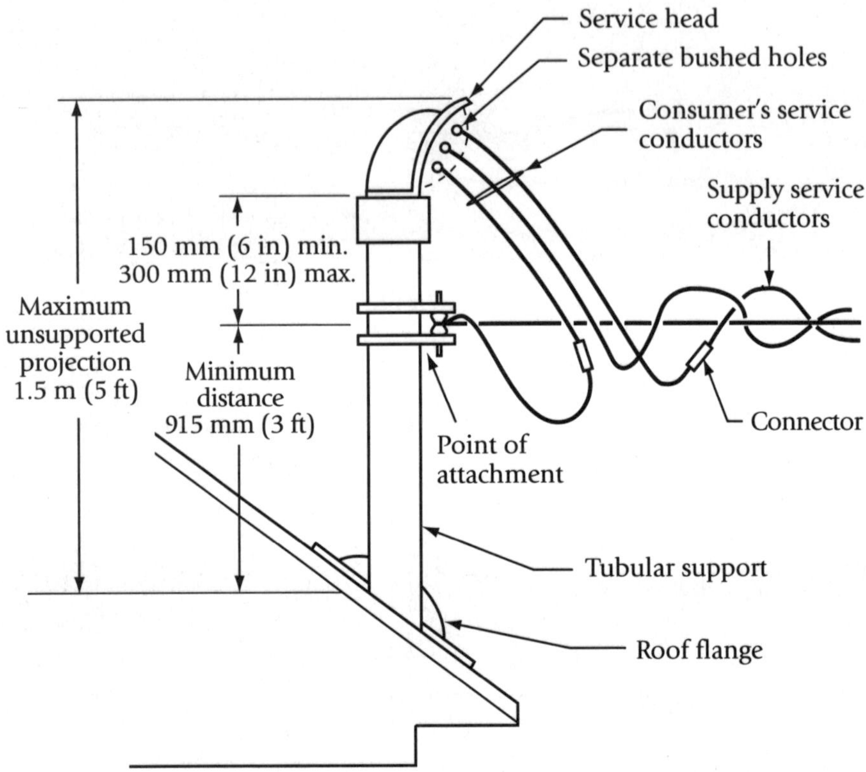


**Figure 6**
**Service head incorporating the point of attachment**

## What are the requirements for a service head?

The service head is the point at which the consumer's service conductors exit the service raceway, the service mast, or the service cable sheath for connection to the supply authority's conductors (see Figures 6 and 7). The service head is also called a weather head, an "F" fitting, or a rain-tight service entrance head.

To comply with the Code, the service head location must be in accordance with any requirements of the supply authority and be located between 150 mm (6 in) and 300 mm (12 in) above the point of attachment of the overhead supply service conductors (the insulator and bracket or rack). The consumer's service conductors must extend at least 750 mm (30 in) out of the service head, pointing downwards and creating drip loops to prevent water from running down the insulation on the conductors and entering the raceway, mast, or cable.

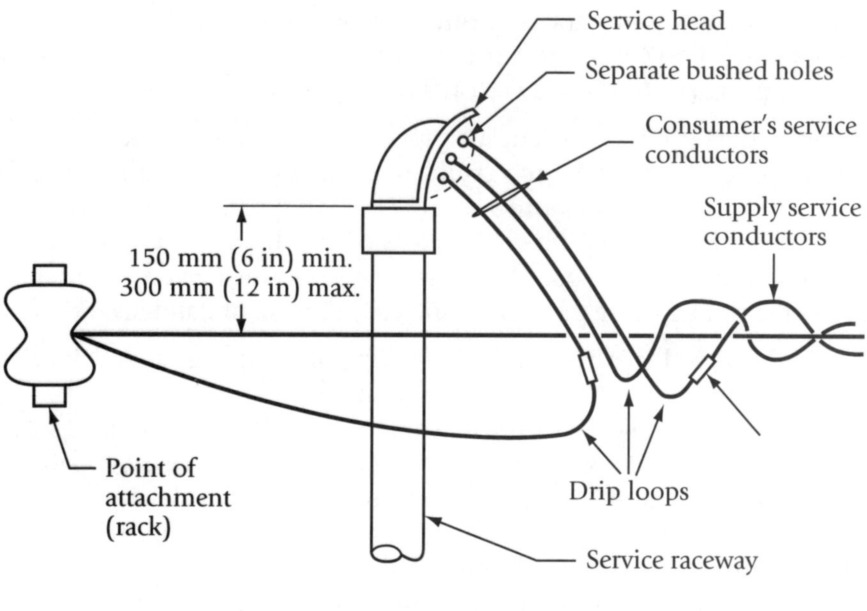


**Figure 7**
**Service head adjacent to the point of attachment**

The insulation on the consumer's service conductors outside of the service head can deteriorate from exposure to weather conditions and environmental pollutants. This may cause the insulation to break down, resulting in short circuits or shock hazards. To reduce the likelihood of these hazards, service conductors of different lines, phases, or polarities are passed through separate bushed holes in the insulating material in the service head fitting. This helps keep the conductors separated, so that even if their insulation deteriorates, they cannot rub together or come into direct contact, causing a short circuit.

## What are the requirements for the service conductors?

The service conductors typically run through a raceway from the service head, along and through the exterior wall, to the service disconnect. The raceway provides protection to the insulated conductors, but it also offers a possible pathway for water to enter the service disconnect from outside. As a result, an above-ground raceway must be drained to the outside and should not enter the top of a service disconnect unless it has been properly drained to the outside.

Raceways are sealed to prevent air movement between the outside and inside of the dwelling, which could lead to condensation inside the raceway and the service equipment. This is especially important in today's dwellings, which have higher-rated insulation levels and more effective vapour and air barriers than older dwellings. Even unfinished basements can be warm enough to increase the risk of condensation.

The Code stipulates that no joints or splices be allowed in the consumer's neutral conductor between the point of connection to the supply authority's service conductors and the line terminals in the consumer's service disconnect. If the continuity of the neutral (white) conductor (see Figure 14 on page 31) breaks down (for example, as a result of an open or high-resistance connection), the line-to-neutral voltage could fluctuate. A change in the line-to-neutral voltage can damage electrical equipment or reduce its life expectancy, as well as create potential shock and fire hazards. However, exceptions are allowed

- if the joint is made by means of a clamp or bolted connection in the meter base; or

- if the joint is made outside the service head used in overhead services at the connection between the supply authority's and the consumer's neutral conductor.

# What are the requirements for underground supply services?

## Where is the service connection?

The broad intent of the Code requirements for underground service installations is much the same as for overhead installations: to prevent accidental contact with the service conductors. In an underground service installation, the service conductors are buried to protect people and machinery from accidental contact with them.

In an underground supply service, the supply authority's conductors and the consumer's service conductors are connected either in a meter base (see Figure 8), usually on the outside wall of the dwelling, or in the service disconnect inside the dwelling. In the meter base, the supply authority's conductors are connected to the line side of the metering equipment. In the service disconnect, the supply authority's conductors are connected to the line terminals of the service disconnect. The consumer's service conductors are connected either to the load terminals of the metering equipment or to the load terminals of the service disconnect.

As with overhead service installations, the conductors running through the service raceway are generally required to be continuous and unbroken, without any joints. However, the Code makes an exception for underground installations

- where the underground portion of an existing underground consumer's service has been damaged;
- where changes are required to accommodate a service relocation; and
- where changes are required to accommodate the relocation of a pole that is part of the service installation.

In these cases, joints may be made in the underground portion, provided that approved underground splice kits are used. It is not the intention, however, to allow joints to be made in a new underground installation, so that, for example, short pieces of cable can be used up, or a piece of cable can be added to make up the correct length of conductor.

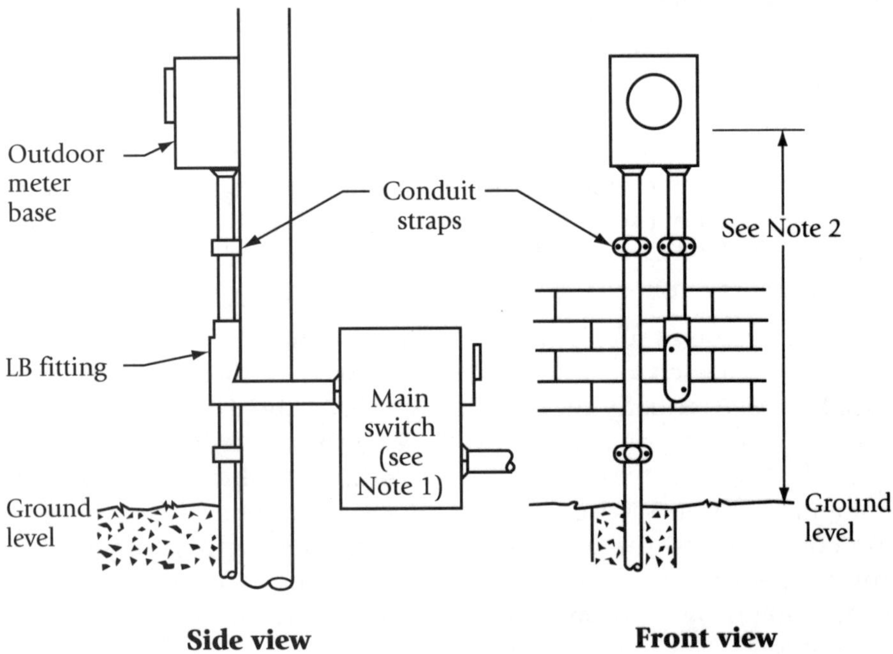


**Notes:**

**(1)** *The meter mounting height is determined by the supply authority and the* National Building Code of Canada.

**(2)** *The main switch may be a combination service disconnect and panel.*

**Figure 8**
**Underground service connection — Meter base**

## How are conductors protected in underground installations?

Service conductors in underground installations are protected by a combination of design features and installation methods. Conductors approved for direct burial usually have thicker insulation that is specially designed to

- withstand the weight of the soil over the conductor; and
- resist deterioration due to moisture and contact with the soil.

On these conductors, the marking "WU" is included in the insulation designation: "W" means the conductor is designed for use in a wet location and "U" means it is designed for direct burial. Underground installations may also use conductors that are not designed for direct burial, by enclosing them in a raceway that is approved for direct burial. Because the raceways bear the weight of the soil, conductors enclosed in such raceways need only be designed for wet locations (marked "W").

## What are the requirements for buried conductors and raceways?

Although the Code contains no specific requirements in this area, the routing of underground supply services (as well as buried feeder or branch circuits, which might supply exterior lighting to a free-standing garage or workshop) is a primary concern. Experience has shown that soil movement due to frost can disturb the buried lines, causing the premature failure of underground conductors. This can occur, for example, where the underground installation runs beneath areas such as roadways, sidewalks, and parking lots, where the snow is removed after a snowfall, as well as beneath areas where snow is allowed to accumulate. The frost level extends deeper in areas where there is no snow cover, and movement of the soil can occur at points where the amount of snow cover varies significantly, putting stress on the buried conductors. Repeated freezing and thawing over the years can cause the conductors' insulation to fail. As a result, potential soil movements and related stresses should be taken into account when selecting the layout and wiring method for an underground installation.

The Code requirements for buried service equipment address

- the minimum depth of the trench or minimum cover;
- the installation of the conductors in the trench;
- the material used to backfill the trench;
- the marking of the underground installation; and
- how the conductor exits the ground.

The minimum trench depth, or more precisely, the minimum depth of cover, is determined by the wiring method, the voltage rating, and whether or not vehicles will move over the area of installation. The following checklist may be used in combination with Table 2 to identify the required depth of cover:

- Do the conductors or cables have a metal sheath or armour?
- Are the conductors or cables installed in a raceway?
- Is the voltage rating above or below 750 V?
- Is it a vehicular or non-vehicular area?

| | **Minimum cover, mm (in)** | | | |
|---|---|---|---|---|
| | **Non-vehicular areas** | | **Vehicular areas** | |
| **Wiring method** | **750 V or less** | **Over 750 V** | **750 V or less** | **Over 750 V** |
| Conductors or cables not having a metal sheath or armour | 600 (24) | 750 (30) | 900 (36) | 1000 (40) |
| Conductors or cables having a metal sheath or armour | 450 (18) | 750 (30) | 600 (24) | 1000 (40) |
| Conductors or cables installed in a raceway | 450 (18) | 750 (30) | 600 (24) | 1000 (40) |

**Note:** *Minimum cover means the distance between the top surface of the conductor, cable, or raceway and the finished ground.*

Table 2

**Minimum cover requirements for direct-buried conductors, cables, or raceways**

When some means of mechanical protection is placed over the conductors, cables, or raceways in the trench, the minimum cover specified in Table 2 may be reduced by 150 mm (6 in). These means include

- treated wooden planking at least 38 mm (1½ in) thick and extending at least 50 mm (2 in) on either side of the conductor;
- poured concrete at least 50 mm (2 in) thick;
- concrete slabs at least 50 mm (2 in) thick and extending at least 50 mm (2 in) on either side of the conductor;
- concrete encasement at least 50 mm (2 in) thick (concrete encasement is usually used only for raceways because the concrete could damage the insulation or sheathing on conductors and cables not specifically approved for concrete encasement); or
- other suitable materials, such as nonmetallic water pipe, nonmetallic drainage ducts, steel pipe, etc., as approved by the local inspection authority.

See Figure 9.

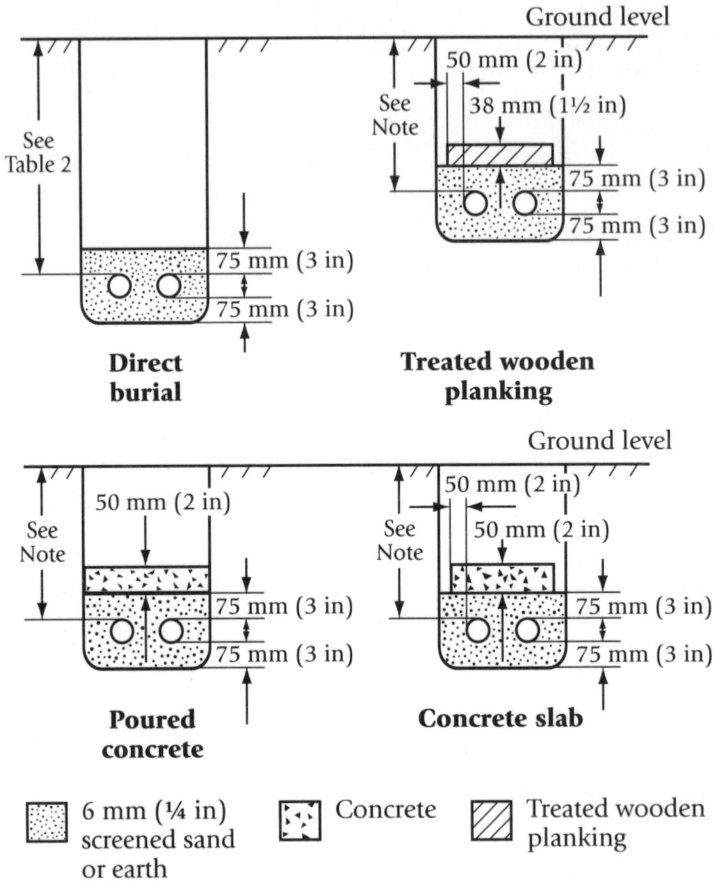


**Note:** *Table 2 requirements reduced by* 150 mm *(6 in).*

**Figure 9**
**Burial options**

Wooden planks that are to be buried in the ground should be treated with a solution of pentachlorophenol or another suitable treatment recommended by a manufacturer of wood preservatives. The use of creosote as a wood preservative is not recommended because it is known to be harmful to rubber and thermoplastic insulations and can therefore cause corrosion of the installation.

If another approved material, such as polyethylene water pipe, is used as mechanical protection for conductors and cables, the conductors and cables must be approved for direct earth burial. Such conductors must have a thicker insulation covering to prevent possible damage, as the inside of a water pipe might not meet the same standards as an approved electrical conduit.

Conductors installed in a trench may be subject to damage if

- sharp or hard objects are forced into the insulation of the conductors or cables; or
- two conductors or cables cross over each other, with the weight of the soil on that point.

The Code requires that a 75 mm (3 in) layer of 6 mm (¼ in) screened sand or earth be placed in the bottom of the trench to prevent sharp or hard objects in the trench from damaging conductors or cables. The conductors or cables are placed on top of the 75 mm (3 in) bedding, running parallel to one another. Conductors must not cross over other conductors or cables along the run in the trench or when entering or exiting the trench. Another 75 mm (3 in) layer of 6 mm (¼ in) screened sand or earth is placed over the conductors and cables, and then the trench can be backfilled. See Figure 9.

Backfill material consisting of large rocks, paving materials, cinders, large or sharply angular substances, corrosive materials, or other materials that could damage the conductors, cables, raceways, or substructures must not be used. As well, the backfill material must not prevent the fill from being adequately compacted. Loosely compacted fill can contribute to the corrosion and damage of the buried equipment through air pockets and the movement of backfill material over time.

The Code requires that underground installations be marked or identified so that people or machines involved in excavation work avoid damaging the installation or coming into potentially dangerous contact with buried conductors. The preferred way of marking the underground installation is to bury a suitable warning marking tape in the trench, halfway between the installation and the finished ground level so that workers will uncover the warning tape well before they come in contact with the underground installation. The other method approved by the Code is to provide adequate marking in a conspicuous location (for example, a signpost placed above the conductors) to indicate the location and depth of the underground installation.

Where conductors or cables enter and exit the trench, the Code requires that they be protected from damage by a rigid conduit terminated vertically in the trench. To prevent damage where the conductors or cables enter or exit the vertical rigid conduit, the Code calls for a bushing, bell-end fitting, or other protective fitting (see Figure 10).

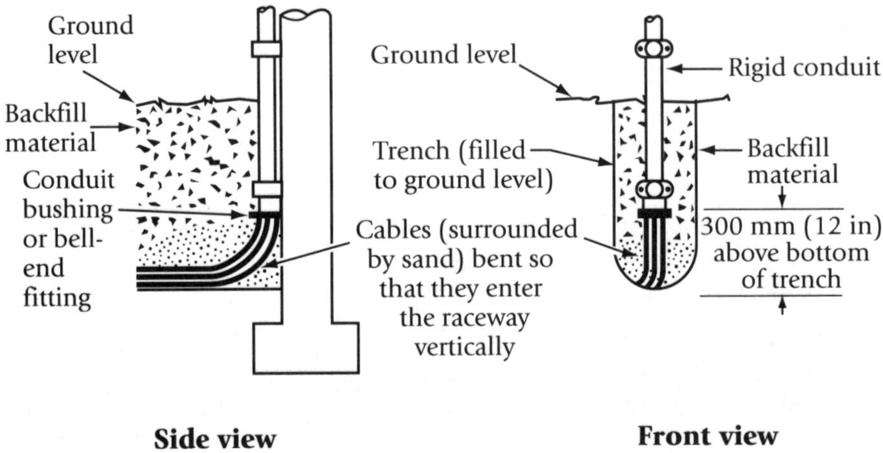


**Figure 10**
**Conductors exiting a trench**

To provide adequate protection for the conductors or cables, the rigid conduit must start 300 mm (12 in) from the bottom of the trench and extend at least 2 m (6 ft) above ground level. The conduit may have to be extended and routed further because of the dwelling's construction features (windows, doors, porches, etc.). There must be enough slack in the conductors or cable where they enter the rigid conduit in the trench so that they enter it in a vertical position, not at an angle. This is intended to prevent any damage that might occur because of the settling of the ground or movement caused by frost.

# What are the location requirements for the consumer's service equipment and metering equipment?

Supply authority personnel, workers, and residential occupants should have a safe and adequate working space when they service, operate, and maintain a consumer's service equipment. The location must be clean and environmentally appropriate for the electrical equipment that is to be installed (see "What are the basic location requirements for a supply service?" on page 4).

The consumer's service conductors, which run from the supply connection at the service head (for overhead supply) or the meter base (for underground supply) to the service disconnect and panel, carry the same load as the supply conductors. The supply authority cannot easily or practically protect the incoming supply conductors because they may be supplying a number of customers.

The consumer's service conductors are unprotected until they enter the service disconnect. They have not been provided with the degree of overcurrent protection that the Code generally requires for conductors located indoors and therefore pose a fire hazard if damaged. As a result, the Code requires that the distance that the consumer's service conductors extend into the building without overcurrent protection be as short as practicable (see Figure 2 on page 7). If this distance is greater than 1.5 m (5 ft), the local inspection department may have special requirements for the installation. The Code permits three wiring methods for these unprotected service conductors. These methods are considered as not presenting a fire hazard, no matter how far the conductors must run before terminating in a service disconnect.

- The service conductors are encased in 50 mm (2 in) of concrete or masonry [see Figure 11(a)].
- The service conductors are located beneath a concrete slab at least 50 mm (2 in) thick [see Figure 11(b)].
- The service conductors are run through a crawl space that is located under the building and that does not exceed 1.8 m (6 ft) in height [see Figure 11(c)].

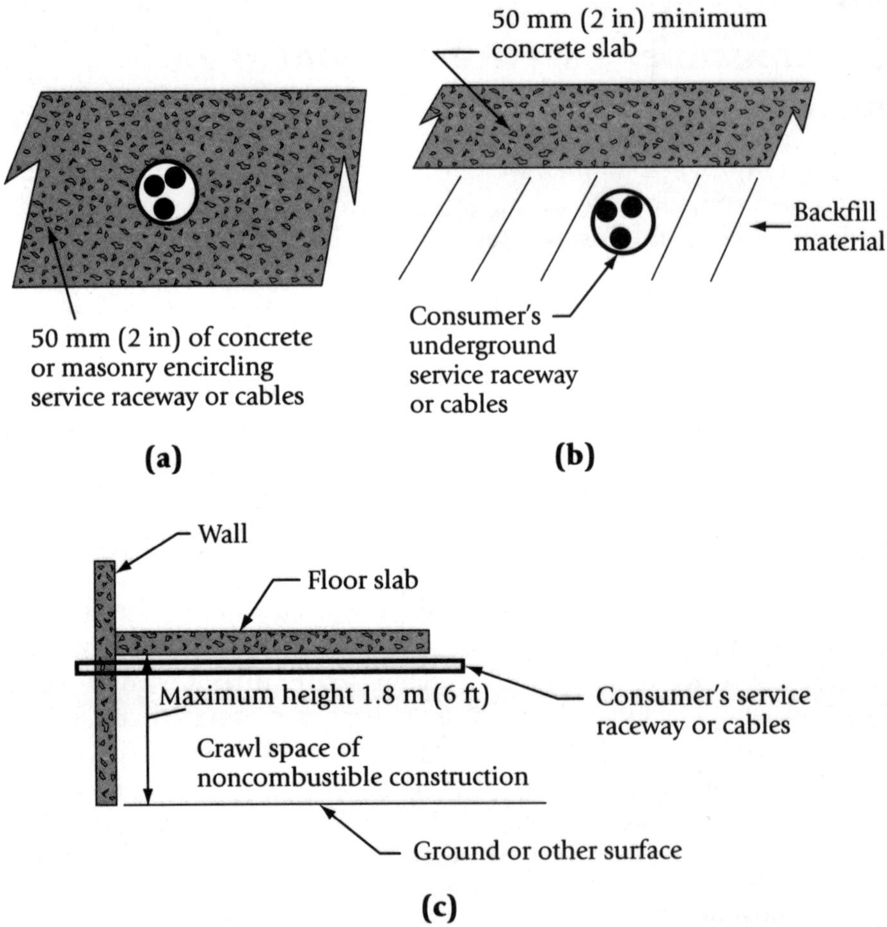


**Figure 11**
**Wiring methods for unprotected service conductors**

The second method (locating the service conductors beneath a concrete slab) is often used where supply authorities require the outdoor service connection to be on the driveway side of a dwelling, for easy access to the meter. Instead of installing the service equipment inside the garage, the conductors are run under the garage concrete slab to the service equipment located in the basement. See Figure 12.

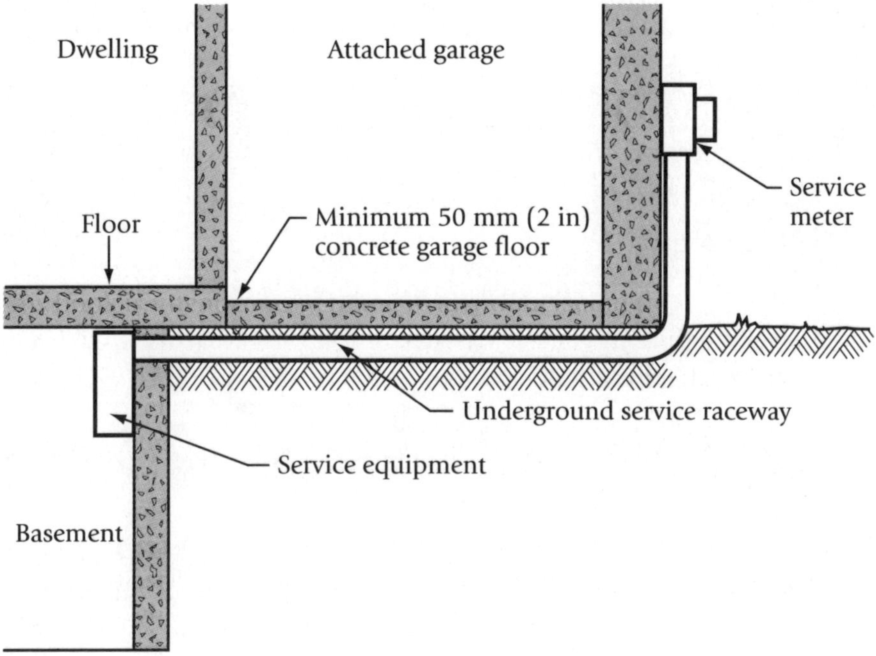


**Figure 12**
**Service conductors installed beneath a garage floor slab**

Where the third method is used (running unprotected service conductors through a crawl space), the crawl space

- must not exceed 1.8 m (6 ft) in height;
- must not be of combustible construction; and
- must not be used for the storage of combustible material.

The rest of the consumer's service equipment, as well as any metering equipment, must satisfy the supply authority's requirements. In general, service equipment must be located inside the building being served, and as close as practicable to the service disconnect, with the exception of meters, which may be located outdoors if they are weatherproof or installed in weatherproof enclosures. If there is more than one meter, they should be grouped together, where practicable. The supply authority may also have special requirements for the location, mounting height, and type of metering equipment.

Service and metering equipment must be accessible without necessitating the use of portable ladders, chairs, etc., or having to climb over or remove obstacles.

Service equipment and meters must not be located in

- coal bins;
- clothes closets;
- bathrooms;
- stairways (and other areas where footing is not level);
- high-temperature areas (near stoves or furnaces);
- areas where the ceiling height is less than 2 m (6 ft); or
- any similar undesirable location.

The Code also requires that when service equipment is installed in a dwelling, the height of the operating handle of any circuit breaker or fuse holder in the panel section not be more than 1.7 m (5.5 ft) from the finished floor level (see Figure 13).

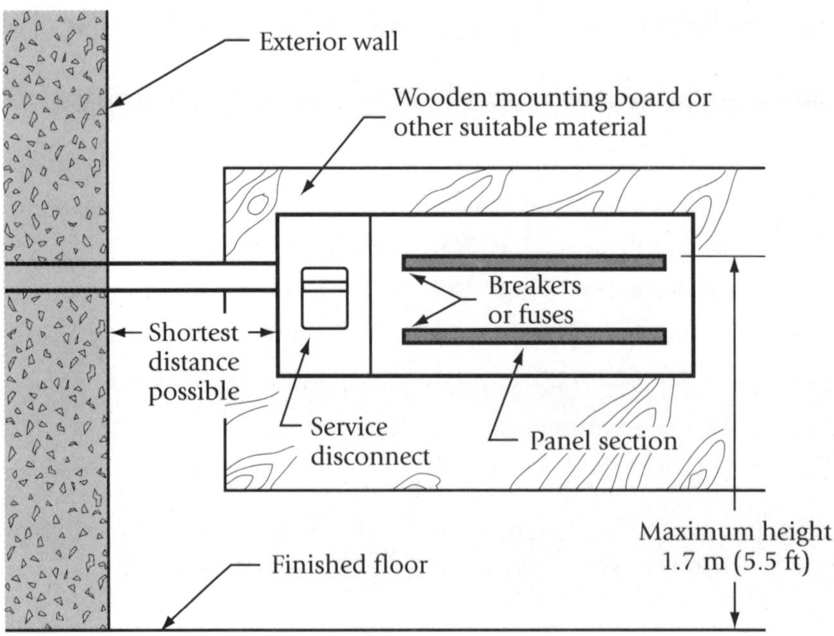


Figure 13
**Installed service equipment**

# 2

# PANELS AND SERVICE SIZE

## What is a panel?

The next important component of the consumer's service is the panel (or panelboard) inside the dwelling. The electrical supply for a panel is controlled by a fusible disconnect switch or a circuit breaker in the feed to the panel. When the disconnect switch or circuit breaker is off (usually for servicing, electrical work, or installation of wiring inside the residence), the electrical supply to the circuits that the panel feeds in the dwelling is effectively shut off. When the disconnect switch or circuit breaker is on, the panel distributes the electrical supply to a number of wiring circuits (or "branch circuits") that extend through the dwelling.

The disconnect switch or circuit breaker should always be turned off before the panel cover is removed and before any work is done inside the panel. A panel should never be left energized without the panel cover on and secured in place.

The panel regulates the load on the dwelling's branch circuits — the amount of current that can flow through the conductors — by means of fuses or, more commonly these days, circuit breakers contained in the panel enclosure. Regulating the load on residential circuits is critical to safety. Electrical circuits and electrical equipment in the home can become hazardous due to short circuits, overloads, and ground faults, which typically result from the deterioration of insulation, the connection of too many loads to an individual circuit, or the failure of a component or components in electrical equipment, appliances, and other devices. The extra load on conductors can cause overheating. If the overload persists, equipment or conductors can reach temperatures that will damage insulation or circuit components and even cause a fire. As well as overloading wiring, these faults can cause arcing, leading to an increased risk of fire and damage to electrical equipment.

Accordingly, the Code requires that all electrical equipment and live (ungrounded) conductors be provided with overcurrent devices (usually fuses or circuit breakers), located in the panel, that will de-energize the circuit if an overload or short circuit occurs. Fuses and circuit breakers sense the current flow going through them in relationship to time. When the current flow, measured in amperes, exceeds the marked rating of the fuse or circuit breaker (the continuous current rating) for a number of seconds, the fuse or circuit breaker de-energizes the circuit that it is protecting. The higher the flow of amperes above the device's marked rating, the sooner it "trips." Circuit breaker overcurrent devices also serve as disconnecting means for the branch circuits, making it possible to cut off electrical supply to just one or two circuits at a time (for servicing, maintenance, changing unswitched bulbs, etc.). Fuses, however, only provide overcurrent protection; they should not be used as a disconnecting means or a switch to energize and de-energize circuits. When fuses are being installed or replaced, the circuit feeding the fuse must be de-energized by opening the disconnecting means or switch that is placed ahead of it in the circuit. If this is not done, arcing can occur when the fuse is inserted or extracted, possibly resulting in injury (burn or shock).

Some circuits are also provided with electrical equipment that will sense the presence of a ground fault and cause the faulted circuit to be de-energized. A ground fault or leakage current is detected by a device called a ground fault circuit interrupter (GFCI) (see Figure 14). The GFCI measures the current flowing out in the circuit and compares it to the current flow returning on the circuit. Any difference in current flow — when less current returns than flows out — is called leakage current: it is "leaking" out of the circuit and, depending on where it goes, could pose a shock hazard. The GFCI trips and de-energizes the circuit when it detects a leakage current of 4 to 6 mA (milliamperes), a level that is not likely to cause major or life-threatening hazards. (See "What are the circuit requirements?" on page 96.)

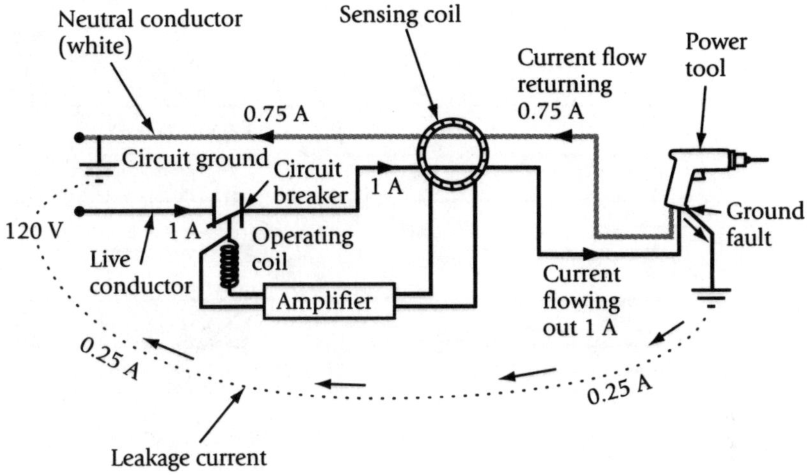


**Figure 14**
**Ground fault circuit interrupter**

In other circuits, the overcurrent device may be supplemented by electrical equipment that will sense the presence of electrical arcs caused by a break in a conductor, the failure of insulation between two conductors, or the breakdown of insulation between a conductor and grounded materials (see Figure 15). An arc fault circuit interrupter (AFCI) recognizes the current or voltage characteristics (or a combination thereof) associated with electrical arcs and trips to de-energize the circuit (see Figure 16 on page 33 and "What are the receptacle and circuit requirements for finished walls in sleeping rooms and areas?" on page 65).

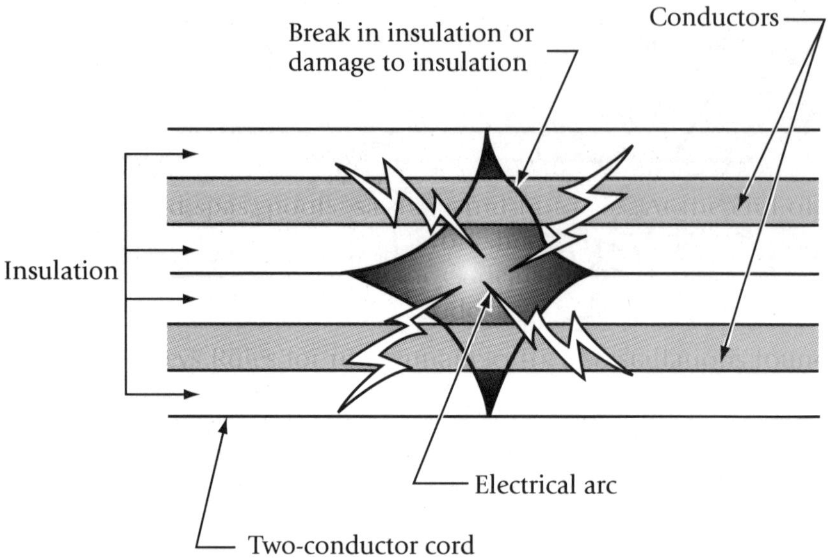


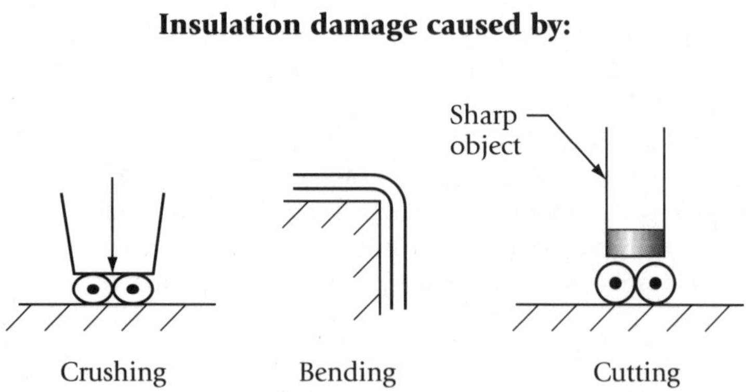


**Figure 15**
**Electrical arcing**

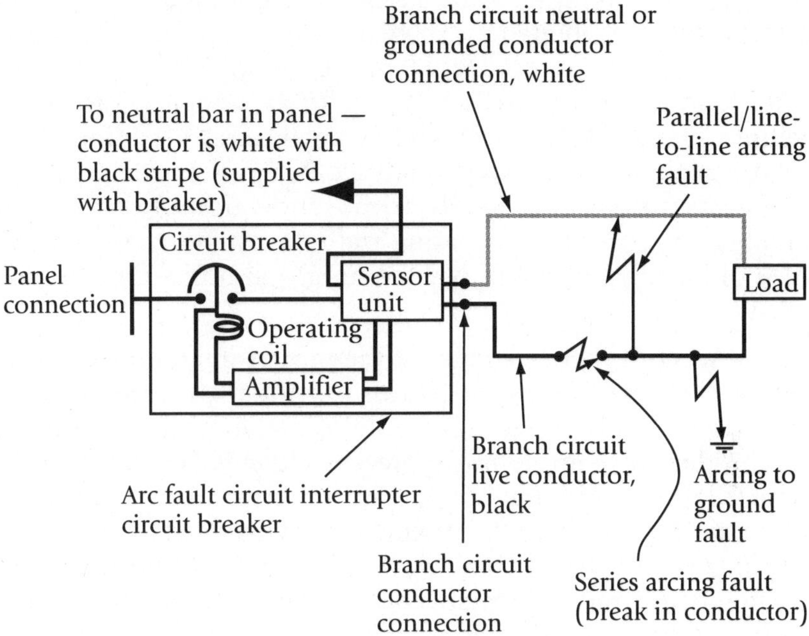


**Figure 16**
**Arc fault circuit interrupter**

## What are the general requirements for panels?

With a few exceptions, the Code requires that each individual dwelling unit be supplied with its own panel. The intention of this requirement is to ensure that the occupants have complete control over all the branch circuits that are being used in their unit. If any particular dwelling unit experiences overloads or faults or requires servicing, the occupants of other dwelling units will not be affected. Similarly, if the supply to the panel must be disconnected — to allow servicing of electrical equipment, for example — other units will not be affected.

Dwelling units that are created by subdividing a single dwelling, and are thus not individually metered, do not require a separate panel. This exception applies only to conversions of an existing single dwelling to create an accessory suite or suites that are not individually metered for power consumption. In many conversions, major modifications to the electrical system are required to ensure that the occupants of the dwelling unit have complete control over all of their branch circuits. Examples include renovating an attic or basement as a small apartment or converting a large dwelling into several small suites. Deciding whether or not to install individual meters and separate panels is largely a question of cost: the price of the electrical renovations (which include cutting, patching, and redecorating walls) must be compared to the benefits of having the tenants pay the supply authority for their use of electricity.

The Code sets out specifications for the provision and installation of panels, while other industry standards provide requirements for their construction. Panels are usually made of metal, both to protect the devices inside from damage and to prevent accidental contact with these devices. If left exposed, the wiring and components of a panel could present a fire or shock hazard. The enclosure is designed to ensure that if a component of the panel fails, any resulting arcing or heating will not pose a fire hazard and will remain contained until an overcurrent device clears the problem. Like other electrical equipment, a panel must be tested and certified to CSA standards. A CSA certification mark (see Figure 17) indicates that the equipment has passed rigorous tests for safety and reliability. Panels must also be properly installed, and because the panel enclosure is made of metal, it must be grounded, so that someone touching any part of the exterior or opening the cover door to access the fuses or circuit breakers will not be exposed to a shock hazard.

*When you buy electrical products, look for the CSA certification mark, a registered trade-mark of CSA.*

**Figure 17**
**CSA certification mark**

## What are the requirements for the location of a panel?

Panels should be located in areas where they will not be subject to damage or abuse, so they are usually installed out of the way of everyday household activities. However, they must also be relatively easy to access and, according to the Code, must be accessible without necessitating the use of portable ladders, chairs, etc., or without having to climb over or remove obstacles.

Panels must not be located in

- coal bins;
- clothes closets;
- bathrooms;
- stairways (and other areas where footing is not level);
- high-temperature areas (near stoves or furnaces);
- areas where the ceiling height is less than 2 m (6 ft); or
- any similar undesirable location.

So, for example, a panel should not be installed in a cupboard or other storage space, where things might be piled up in front of it, making access difficult.

The location of the panel must comply with the requirements of the Code as well as any local supply authority requirements. Consideration should also be given to the possible future uses of the area where the panel is installed, and whether it would interfere with these uses. Moving a panel can be both expensive and inconvenient.

Panels are usually installed as high as possible, so that they can be accessed by adults but kept out of the reach of small children. However, the Code stipulates a maximum height, requiring that any operating handle for an overcurrent device be no more than 1.7 m (5.5 ft) above the finished floor. This maximum height is meant to allow convenient access and helps to ensure that people will be able to read the panel directory and any warning notices that may be displayed on the panel. (In dwellings having barrier-free design, panel heights may vary. The *National Building Code of Canada* and local codes should be consulted for the maximum mounting height of the panel. CSA Standard CAN/CSA-B651, *Barrier-Free Design*, also provides guidance.)

Circuit breaker panels can be mounted in any direction (vertical or horizontal). This is usually determined by the number and the direction of cables that will enter the panel, and where the knockouts are situated along the ends of the panel. The panel is attached to a backboard, which can be used to support the cables entering the panel (see Figure 18). The Code requires that the panel directory, usually a set of labels adjacent to the circuit breaker positions, be filled out completely, so that the dwelling's occupants, as well as anyone servicing electrical equipment, know which overcurrent device protects and controls each circuit.

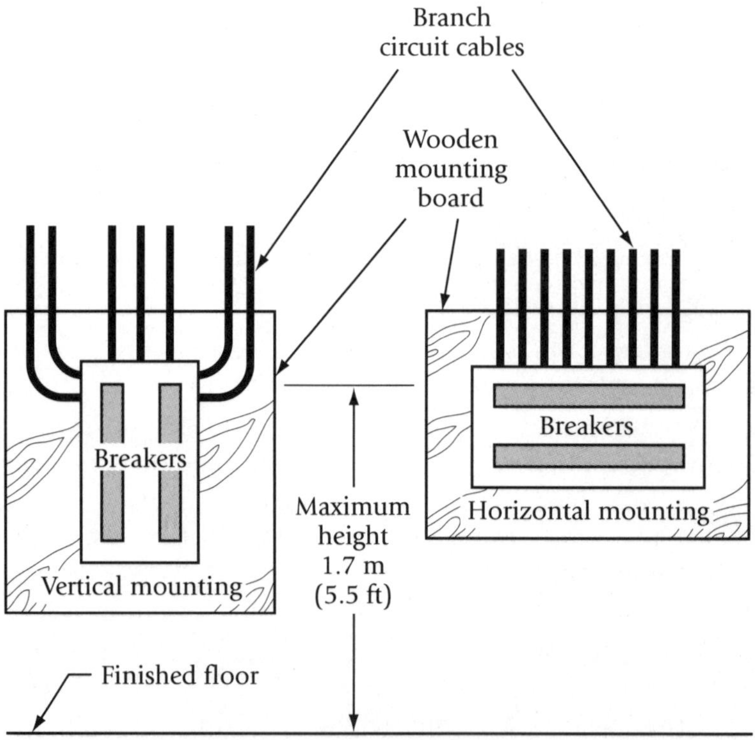


**Figure 18**
**Mounted panel**

# What are the requirements for the overcurrent protection of panels?

The panel itself is provided with overcurrent protection, either as an integral part or as an overcurrent device on the supply side of the panel, to protect the panel as a whole from overloading or short circuits. The rating of the panel's overcurrent device depends on the marked rating of the panel. The Code requires that the overcurrent device for a panel not exceed the marked rating of the panel. That rating and the number of branch circuit positions are determined by the loads that the overcurrent devices installed in the panel are expected to supply. The number of loads to be supplied depends, in turn, on the number and intended uses of the branch circuits (i.e., the size of service; see "What is the size of service?" on page 40). In residential installations, a panel typically supplies a combination of lighting, heating, appliance, and receptacle branch circuits. It may also supply motor circuits (for example, a furnace, water and/or sump pump, central air conditioner, etc.) and feeder circuits — circuits that supply power to other, usually smaller, panels — for example, a panel in a workshop, a detached garage, or a pool cabana. It is generally a good idea, and often cheaper in the long run, to install feeder circuits and smaller panels in such areas, to serve existing needs and meet heavier future requirements.

Circuits controlled by the panel may supply a number of loads that are not all expected to be in use at the same time. This diversification makes it possible to use a panel protected by an overcurrent device that is rated at *less* than the sum of the ratings of all the overcurrent devices installed in the panel or the rating of the panel. For example, a 125 A panel might be protected by a 60 A overcurrent device, or a 200 A panel might be protected by a 100 A overcurrent device, so long as the total load on the panel at any time does not exceed the rating of the overcurrent device protecting the panel. If the load at any time does exceed that rating, the overcurrent device will trip, de-energizing the panel and cutting off power to all the circuits that the panel supplies. Using a panel with a higher rating makes more branch circuit overcurrent device positions available, making it easier to reduce the load on an individual branch circuit by adding another circuit, or to expand the electrical service at a later date.

## What are the requirements for branch circuit positions in the panel?

The Code stipulates a minimum number of branch circuit positions that corresponds to the rating of the panel in amperes (see Table 3). While the minimum number is acceptable, it is good practice to install a panel with more than the minimum number of branch circuit spaces required. This makes it easier to reduce the load on existing individual branch circuits and to add new circuits as electrical requirements change.

| Rating of panel, A | Minimum number of branch circuit positions |
|---|---|
| 60 | 16 |
| 100 | 24 |
| 125 | 30 |
| 200 | 40 |

**Table 3**
**Required number of branch circuit positions**

At least half of the branch circuit positions must accommodate a double-pole (connected to two different lines) overcurrent device to serve general-purpose branch circuits. The panel must also include space for two 35 A double-pole overcurrent devices, for circuits that may be needed to serve electric ranges, clothes dryers, central air conditioners, some power tools, secondary panels, etc.

As well, at least two additional positions must be left for future use. In the past, when a panel had all its branch circuit positions filled, an extra circuit controlled by a fusible switch, or a smaller secondary panel, was tapped off the mains of the existing panel (see Figure 19). The smaller conductors required for the new switch or panel were connected under the same lug as the supply conductors for the existing panel. This practice created some problems: the switch or panel tapped off the mains of the existing panel often were not properly protected, and the lugs in the existing panel were not usually designed, tested, or approved for two conductors, especially conductors of different sizes.

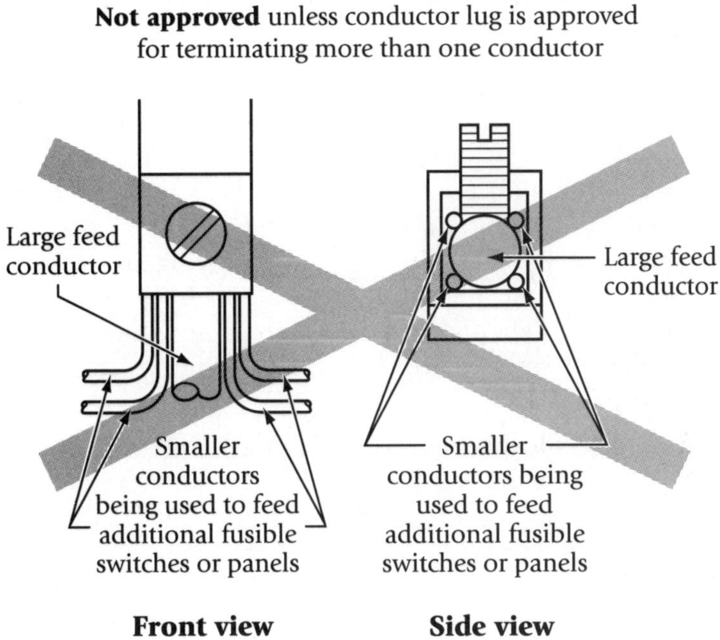


Figure 19
**Conductor taps**

Leaving at least two branch circuit positions for future expansion makes it possible to provide proper protection for the conductors that feed additional panels and makes it unnecessary to try to connect two conductors in a lug that is not designed and tested for that purpose. When expanding the electrical service, it is good practice not to use a spare position in a panel to add an individual circuit, but to use it to feed a small panel. Doing so will supply extra branch circuit positions, ensure proper protection for the panel board, and maintain two spare positions for future use.

## What is the size of service?

Size of service refers to the maximum amount of electricity, measured in amperes (A), that a panel and its overcurrent devices can supply to residential circuits. It is basically determined by a dwelling's area and the electrical demand, measured in watts (W), of the dwelling's electrical circuits and equipment. In the 1920s, 30 A installations were common, but the Code now stipulates a minimum service size of 60 A for single dwellings, with the added condition that the living area of the ground floor is less than 80 $m^2$. Local authorities may have requirements for a larger service. As a result, most new homes now have 100 or 200 A services, and larger services are often installed in anticipation of high demands for power.

The service for a single dwelling is sized to meet the existing loads in the dwelling and also to allow for new loads that might be added, such as air conditioning, a spa or hot tub, a swimming pool, and new wiring for an addition or renovation, a finished basement, a workshop, or an outbuilding (for example, a garage or storage shed). Cost is also a factor in the decision because it is usually cheaper to install a service with a larger capacity than required at present than to upgrade the service at a later date.

## How is the size of service calculated for a single dwelling?

To calculate the size of a new service, or to assess the adequacy of an existing service, you need the following information:

- the area of the ground floor, based on the inside dimensions (in square metres);
- the area of the basement, based on the inside dimensions (in square metres);
- the area of any living area on floors other than the basement and ground floor, based on inside dimensions (in square metres);
- the total of any electric space-heating loads (in watts);
- the total of any air-conditioning loads (in watts);
- the size of any electric range or stove loads (in watts);

- any electric tankless water heater loads (in watts);
- any electric water heater loads for in-line water heaters (not the storage type), steamers, spas, hot tubs, and swimming pools (in watts);
- any additional loads with a rating over 1500 W (for example, a clothes dryer, an electric storage-type water heater, and similar loads that are direct-connected, i.e., not plugged into a general-purpose receptacle); and
- any load over 1500 W that is expected to be added in the future.

Measuring floor areas is usually simple, but identifying the load rating of electrical equipment may require some time and ingenuity. Most equipment with high-load ratings (more than 1500 W) will have markings, nameplates, tags, manufacturer's brochures, or manuals that supply the required information. Appliances and equipment with ratings of 1500 W or less need not be included individually in the list, as these basic loads are accounted for in the calculation by adding a standard number of watts per unit of area (see Step 2 below).

The size of service is calculated as follows (a worksheet is provided on page 183):

**Step 1** Determine the total living area of the dwelling in square metres:

ground floor area + (basement floor area × 0.75) + floor area of any living areas other than the basement and ground floor = total living area, $m^2$

If a single dwelling has an attached garage, include the floor area of the garage as part of the ground floor area.

**Step 2** For the first 90 $m^2$ of the total living area, allocate 5000 W. Add another 1000 W for each additional 90 $m^2$ of living area and 1000 W for any leftover portion less than 90 $m^2$. (This accounts for all anticipated basic loads.)

**Step 3** Determine the electric space-heating load, if any, calculated as an initial 10 kW of the rating plus 75% of any remaining load, according to the equipment's rating. (For example, the load for electric space heaters having a total rating of 20 kW would be 10 kW + (10 kW × 0.75) = 17.5 kW.)

**Step 4** Determine the air-conditioning load, calculated as 100% of the equipment's rating.

**Step 5** To the basic load determined in Step 2, add the larger of the two loads calculated in Steps 3 and 4. (It is assumed that the electric space-heating load and the air-conditioning load will not occur at the same time.)

**Step 6** Add the electric range load, calculated as 6000 W for the first 12 kW of the rating plus 40% of any remaining load.

For example, the load for a 15 kW range would be
6000 W + (3000 W × 0.4) = 7200 W

If the rating of the electric range is not known, assume that the rating is 12 kW, and add 6000 W.

**Step 7** Add 100% of the ratings of any electric tankless water heaters, electric steamers, spas, hot tubs, or electric swimming pool heaters.

**Step 8** If the dwelling has an electric range load (see Step 6), then add 25% of each individual additional load greater than 1500 W. If the dwelling does not have an electric range load, add 100% of each individual additional load greater than 1500 W, up to 6000 W, and 25% of any remaining individual load in excess of 1500 W.

**Step 9** Add 100% of any anticipated future loads.

**Step 10** Select the appropriate size of service from Table 4, according to the calculated total loads and taking into account the need to make allowance for future loads.

| Total load, W | Size of service, A |
|---|---|
| 14 400 | 60 |
| 24 000 | 100 |
| 30 000 | 125 |
| 36 000 | 150 |
| 48 000 | 200 |
| 54 000 | 225 |
| 96 000 | 400 |

**Table 4**
**Calculating size of service**

## How is the size of service calculated for a multiple dwelling?

More than one consumer's service may be connected to a single supply service, for example, in duplex, triplex, and quadruplex installations, or in a dwelling that requires a smaller dwelling unit within it to be metered separately from the meter for the single supply service. In most of these cases, there is not a main service disconnect switch for the complex; instead, each dwelling unit has its own service equipment, fed from a meter base that accommodates multiple meters. The meters are usually placed outdoors, ahead of the individual dwelling unit's service equipment (see Figure 1 on page 2 and Figure 2 on page 7).

Whether there is one main service disconnect switch or multiple meters, the method below is used to size the service. This method may be used for two or more dwelling units connected to a single service; where there are more than ten, the Code provides more specific and detailed requirements for the sizing of service equipment.

For between two and ten dwelling units connected to a single service, the size of service is calculated as follows (a worksheet is provided on page 189):

**Step 1** Using the method set out in "How is the size of service calculated for a single dwelling?" on page 40, determine the size of service in watts required for each dwelling unit, but do not include any electric space-heating loads or air-conditioning loads.

**Step 2** Determine which dwelling unit has the largest load in watts.

**Step 3** Determine the two units with the next largest loads. Multiply those loads by 0.65 and add the value to the load identified in Step 2.

**Step 4** Determine the two units with the next largest loads. Multiply the loads by 0.4 and add the value to the running total.

**Step 5** Determine the five units with the next largest loads. Multiply the loads by 0.25, and add the value to the running total.

**Step 6** Determine the total electric space-heating loads in all the dwelling units, calculated as the initial 10 kW of the rating plus 75% of any remaining load.

**Step 7** Determine the total air-conditioning load in all the dwelling units, taken as 100% of the equipment rating.

**Step 8** Add the larger of the two loads calculated in Steps 6 and 7 to the load determined in Steps 3 to 5. (It is assumed that the electric space-heating load and the air-conditioning load will not occur at the same time.)

**Step 9** Add all the loads outside of the individual dwelling units (these loads may be described as common loads used by all the dwelling units), multiply by 0.94, and add the load to the overall total load.

**Step 10** Select the appropriate service size from Table 4.

# 3

# Branch circuits

## What kinds of branch circuits are found in residential installations?

Typically, the panel distributes electrical power through four kinds of circuits:

- general-purpose circuits;
- multi-wire (or split) circuits;
- 20 A T-slot receptacle circuits; and
- feeder circuits.

The Code also provides requirements for emergency, fiber optic, and transformer circuits. Though these are seldom found in residential installations, in cases such as the addition of communication, monitoring, or alarm systems, or the installation of industrial-quality power tools, the Code should be consulted for any applicable requirements.

## What is a general-purpose circuit?

A general-purpose circuit is a two-wire branch circuit that supplies current to a number of outlets that are rated 1 A or less. These low-power outlets, often called standard outlets, are used for the installation of ordinary lighting fixtures and the general-use receptacles that are provided in the finished walls of a dwelling. (The word "outlet" is commonly used to refer to a general-use receptacle, but "outlet" actually covers any device that is installed in a branch circuit to deliver power to electrical equipment.) General-purpose circuits can be used to supply lighting outlets only, receptacle outlets only, or a combination of receptacle and lighting outlets. According to the Code, a general-purpose circuit may have up to a maximum of 12 such outlets. Switches (for example, single-pole or 3-way switches) are not considered to be outlets and are not included in the number of outlets connected to a two-wire circuit.

## What are the general requirements for general-purpose circuits?

Although the Code allows a maximum of 12 outlets on any one general-purpose circuit, it is often prudent to have fewer outlets, and if necessary more circuits, so as to reduce the likelihood of accidental overloads and nuisance tripping of the fuse or circuit breaker protecting the circuit. For example, the receptacle that is required next to the sink in a bathroom or powder room might be used for appliances such as hair dryers or curling irons, which draw a relatively high current. Receptacles in an office area could supply a computer, printer, scanner, fax machine, copier, coffee maker, etc. The combination of these loads and other lower loads on the outlets on the circuit could cause nuisance tripping of the fuse or circuit breaker protecting the circuit. As a result, a good practical rule is to place a maximum of 10 standard outlets on any general-purpose circuit. This allows the circuit some extra capacity to accommodate intermittent high loads and also provides for the possibility that another outlet might have to be added to the circuit at some future point.

Having fewer outlets and more circuits can also prevent potentially dangerous conditions when fuses or circuit breakers do trip. When only one circuit is used to supply all the lighting and receptacles in an area — as is often the case in bathrooms and basement recreation rooms — the tripping of the circuit's overcurrent device leaves the area in total darkness, creating a safety hazard. In the event of a power outage on the supply authority's system, this sort of hazard will still arise, but a well-designed circuit layout can at least anticipate outages on individual circuits.

## What is a multi-wire (split) circuit?

A multi-wire circuit, or split circuit, is a three-conductor circuit consisting of a live (ungrounded) conductor from each line and a neutral conductor. The three conductors are the same size. A multi-wire circuit is essentially two circuits that share a common neutral (a conductor that usually has white insulation or is otherwise identified as the neutral), instead of each having its own neutral conductor. In this type of circuit, the neutral conductor carries the unbalanced load, which is the difference between the loads on the two conductors. In a general-purpose, two-conductor branch circuit, the line conductor and the neutral conductor both carry the same load.

Multi-wire circuits are commonly installed to supply areas where two appliances or devices that have high ratings (usually 1000 W or higher) are to be connected to the same outlet box or plugged into the same receptacle. So, for example, a multi-wire circuit is used to supply receptacles for kitchen counter work surfaces and for power tools and equipment in workshops. A multi-wire circuit might also be used to supply a lighting circuit and a circuit for receptacles in a basement recreation room.

## What are the general requirements for a multi-wire (split) circuit?

When installing a multi-wire circuit and connecting it to the panel, it is important that each line conductor be connected to a different line in the panel. If the two conductors were connected to the same line, the neutral conductor would have to carry the sum of the loads in both line conductors, instead of the difference, and could overheat, causing a fire hazard.

To help prevent incorrect installations, multi-wire circuits are usually connected to a 240 V fuse connection block in a fuse panel or to a double-pole circuit breaker in a circuit breaker panel. In some cases, however, the Code permits two individual fuses or circuit breakers to be used, so long as the multi-wire circuit is not connected to just one device. For example, a multi-wire circuit that uses one line to supply lighting fixtures and the other line to supply one or more receptacles for power tools in a workshop would qualify for this exception. A multi-wire circuit that supplies only kitchen counter receptacles would not qualify because this could create a shock hazard. A homeowner might remove one of the fuses or turn off one of the breakers, thinking that this would shut off the circuit's power at the device. In fact, the power would not be turned off completely because the other line conductor to the device would still be connected, and the neutral conductor would still be working as the return conductor for that circuit. Accordingly, where circuit breakers are used, inspection authorities require that the handles of the two circuit breakers be tied together by a handle tie supplied by the manufacturer. Some authorities insist that a double-pole circuit breaker be used for any multi-wire circuit, even where each live conductor is connected to a separate outlet or device.

Electricians and electrical inspectors use a voltmeter to test whether a multi-wire circuit has been connected correctly. This test is conducted either on the live electrical panel, with the cover off, or at the energized multi-wire receptacle. It should be performed only by persons who have been trained in using voltmeters to test energized or live circuits with exposed live terminations.

## What are 20 A T-slot receptacles (small appliance branch circuits)?

The 20 A T-slot receptacle gets its name from the shape of the slots for the neutral blades of a portable male cord/plug cap. The shape of the slot for the neutral connection is the capital letter "T" on its side. This type of receptacle can accommodate male cord/plug caps at a rating of either 15 A at 120 V or 20 A at 120 V. The 20 A T-slot receptacle is used to supply high-wattage portable appliances, such as kitchen appliances, central vacuum units, and window air conditioners, and workshop receptacles, laundry room receptacles, and garage and outdoor receptacles.

## What are the requirements for 20 A T-slot receptacles?

The Code allows 20 A T-slot receptacles to be installed in any room or area in a dwelling. The minimum conductor size is #12 AWG, and the circuit must have an ampacity rating of 20 A and be protected by a 20 A maximum overcurrent device. Other Code requirements may apply in connection with ground fault protection, the maximum number of receptacles on a circuit, etc.

## What is a feeder circuit?

A feeder circuit runs from the main service equipment to supply another, usually smaller panel that contains overcurrent devices. Feeders and secondary panels are usually used to supply loads in specific areas, such as detached garages, workshops, basements, second kitchens, granny suites, etc.

In a feeder circuit, there is a set of overcurrent devices at the supply end and at the load end. The electrical characteristics of the secondary panel, as designed and constructed, determine

- the maximum rating of the overcurrent device that protects the panel; and
- the minimum size of the conductor used to feed the panel.

These characteristics are provided on a panel's nameplate. A panel may be protected by an overcurrent device with a lower rating than the panel's maximum rating. The primary reason for using a panel with an ampacity rating higher than the rating of the overcurrent device is that such a panel provides more branch circuit positions. These added positions might be required because of the type of circuits being installed or to make provisions for planned expansions. Also, where the number of outlets placed on general-purpose branch circuits is reduced, having more branch circuit positions allows more branch circuits to be run, as long as there is no overall increase in load (that is, not all the loads are expected to be used at the same time). With more circuits and lower loads on each individual circuit, there is less chance of nuisance tripping of individual circuit breakers, should higher loads be placed on the circuits. In some cases, for economic reasons, people install a large panel and change the service size to match the panel at a later date.

## What are the general requirements for feeder circuits?

The conductor used to feed the panel must not have a lower rating than the overcurrent device itself, regardless of the rating of the panel. For example, a panel rated for 200 A might be protected by a 100 A overcurrent device; in this case, the minimum ampacity of the conductor feeding the panel would have to be 100 A. Table 5 shows how to size the conductors for a feeder that supplies a panel: identify the rating of the panel's overcurrent device, and select the conductor that has at least that ampacity rating.

## How are branch circuits sized?

The Code provides specific overcurrent protection and conductor size requirements for most of the branch circuits required for various loads in dwellings. On occasion, however, it may be necessary to calculate the minimum size of conductor for a circuit and the maximum rating of the overcurrent device that must be used to protect the circuit (for example, to assess the adequacy of existing wiring or to plan for renovations or for the installation of new appliances, devices, or equipment). The calculations are made as follows:

**Step 1** From the nameplate on the electrical equipment, determine

- the total load in watts;
- the voltage rating of the equipment; and
- the ampere rating of the equipment (if given).

If the ampere rating of the equipment is not shown on the nameplate, it can be calculated by dividing the total wattage by the voltage rating of the equipment:

Amperes = Watts/Volts(A = W/V)

**Step 2** The minimum ampacity of the individual branch circuit conductor is calculated as 125% of the ampacity of the equipment:

Minimum conductor ampacity = Equipment ampacity × 1.25

**Step 3** Determine the minimum size of branch circuit conductor from Table 5.

**Step 4** Determine the maximum rating of the branch circuit overcurrent device (fuse or circuit breaker) from Table 5. The maximum rating of the overcurrent device must not be greater than the ampacity of the branch circuit conductor.

For example, a homeowner might intend to install a sauna with an electric heater rated at 4.5 kW, 240 V, 18.8 A. The circuit would be sized as follows:

**Step 1** The rating of the heater in amperes is 18.8 (given by the manufacturer).

**Step 2** The equipment ampere rating is multiplied by 1.25:

18.8 × 1.25 = 23.5 A

**Step 3** From Table 5, the required size of conductor is #10 AWG.

**Step 4** From Table 5, the required size of overcurrent device (fuse or circuit breaker) is 30 A.

| Conductor ampacity | Conductor size, AWG | Maximum overcurrent device rating, A |
|---|---|---|
| 0–15 | #14 | 15 |
| 16–20 | #12 | 20 |
| 21–30 | #10 | 30 |
| 31–45 | #8 | 40 |
| 46–65 | #6 | 60 |
| 66–85 | #4 | 80 |
| 86–100 | #3 | 100 |
| 101–115 | #2 | 110 |
| 116–130 | #1 | 125 |
| 131–150 | #0 | 150 |
| 151–175 | #00 | 175 |
| 176–200 | #000 | 200 |
| 201–230 | #0000 | 225 |

**Table 5**
**Sizing circuits**

# 4

# RECEPTACLES

## What are the general requirements for receptacles?

In the earliest residential electrical installations, electricity was used only to provide fixed lighting; there was no need for receptacles. As various types of electrical appliances and fixtures came onto the market, it became common to install receptacles. For some time, it was thought that providing one receptacle in every room was sufficient. However, the receptacle was frequently not in the location where it was needed, and extension cords became a must. Although it was common to read in fire reports that fires were caused by faulty electrical wiring, many such cases actually involved the misuse of extension cords. Extension cords, often undersized, were run through closets and cupboards, behind baseboards, under carpets and linoleum, and even under wallpaper. There was clearly a need for more receptacles in dwellings, to reduce the use of extension cords and to make the use of electrical appliances and fixtures safer.

Since 1927, the Code has set out an increasingly detailed set of rules for the provision and location of receptacles. These rules differ according to the area in which receptacles are being installed, taking into account construction features, intended uses, and special circumstances that might pose a hazard. This chapter considers the most common locations for receptacles and reviews the most important requirements in each case. For an "at a glance" summary of receptacle requirements, see "Receptacle requirements at a glance" on page 171.

# What are the requirements for receptacles in finished walls?

## What is a finished wall?

The Code requires a minimum number of duplex receptacles along the perimeter of a finished wall. A finished wall is any wall finished, from ceiling to within 450 mm of the floor, with drywall, wood, other types of panelling, or finishing materials such as plaster. The term applies to most interior walls.

The Code also makes exceptions for certain rooms or areas in a dwelling if the use or function of the room or area makes it unnecessary to specify a minimum number of receptacles. According to the Code, some of the rooms and areas in a dwelling — kitchens, bathrooms, hallways, laundry rooms, utility rooms, and closets — are not considered to have finished walls and need not comply with the general requirements for the provision of duplex receptacles. (Special requirements for these rooms and areas are discussed later in this chapter.)

The Code also makes an important distinction between ordinary rooms and rooms used for sleeping and sets out special circuit requirements for receptacles installed in the finished walls of sleeping rooms and areas (see "What are the receptacle and circuit requirements for finished walls in sleeping rooms and areas?" on page 65).

## What is usable wall space?

The required number of receptacles depends on the usable wall space of a finished wall. Usable wall space is any wall space 900 mm (36 in) or more in width, measured along the floor line of the finished wall. Usable wall space does not include doorways, areas occupied by a door when the door is fully opened, windows that extend to the floor or to within 300 mm (12 in) of the floor (such as most standard dimension bow windows), fireplaces, or other permanent installations (such as bars, cabinets along the floor, and permanent bookshelves) that would limit the use of the wall space. See Figure 20.

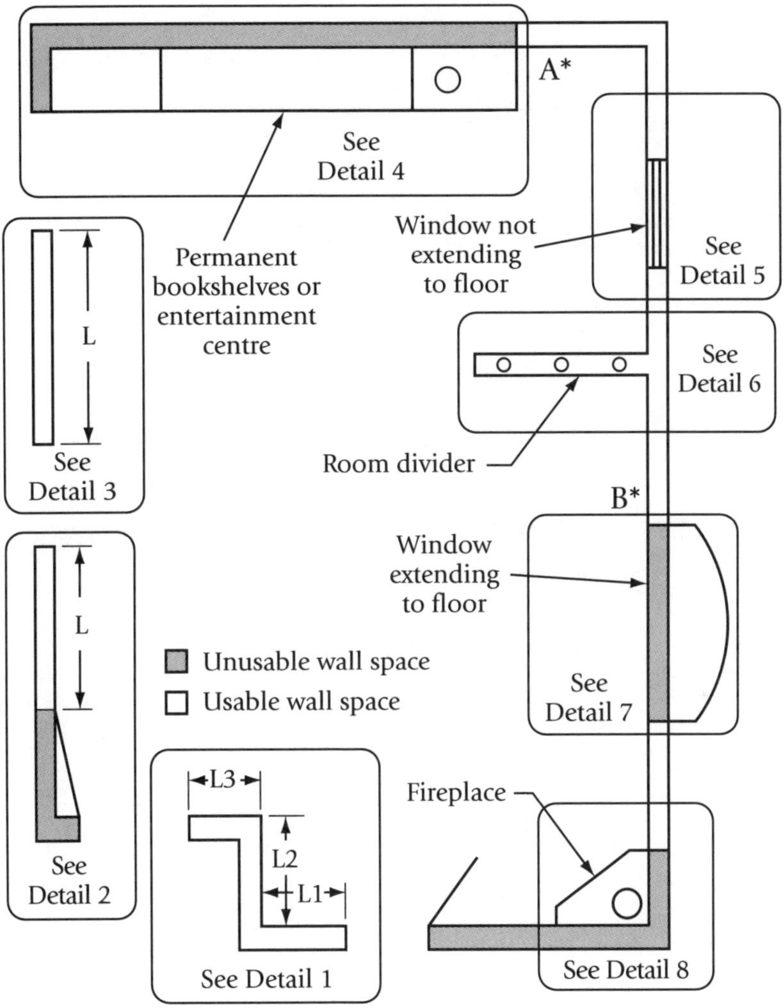


** Where the distance between point A and point B is greater than* 900 mm *(36 in), measured along the floor line of the wall, the wall space is considered usable.*

**Figure 20**
**Usable wall space**

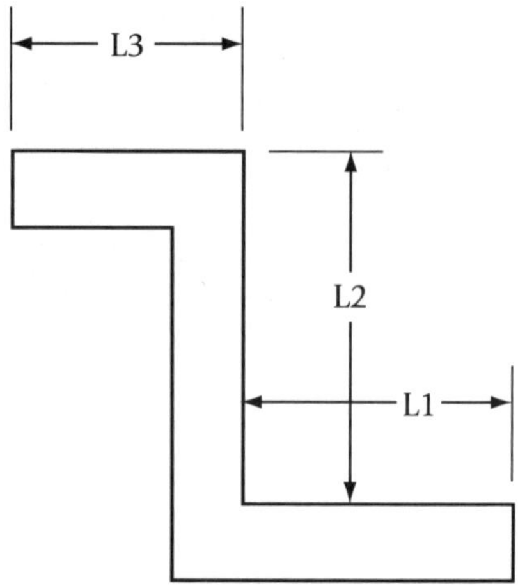


L1 + L2 + L3 = 900 mm (3 ft) or more — usable wall space

**Figure 20 — Detail 1**

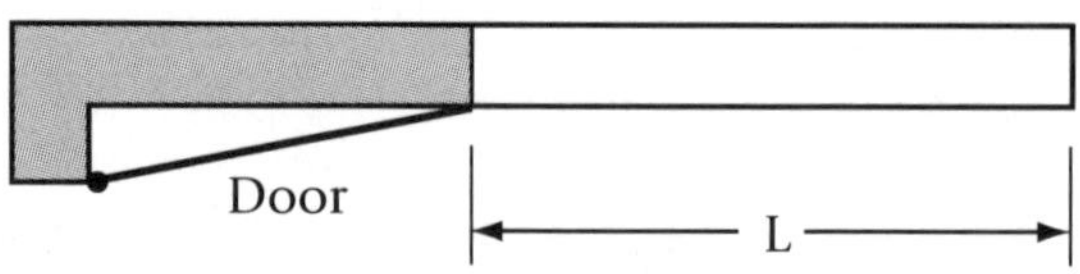


Length of wall behind door — unusable wall space
L is less than 900 mm (3 ft) — unusable wall space

**Figure 20 — Detail 2**

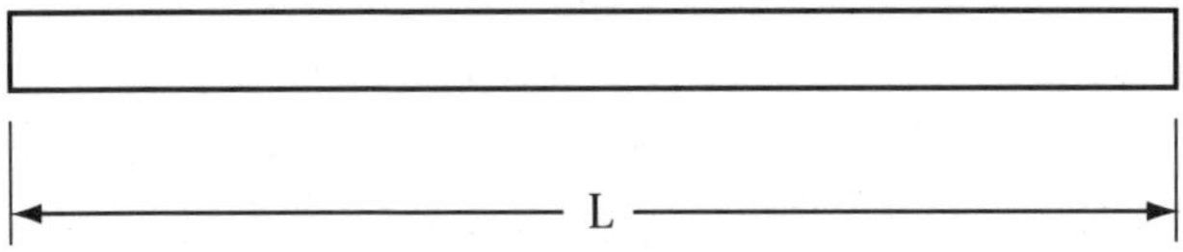


L is 900 mm (3 ft) or more — usable wall space

**Figure 20 — Detail 3**

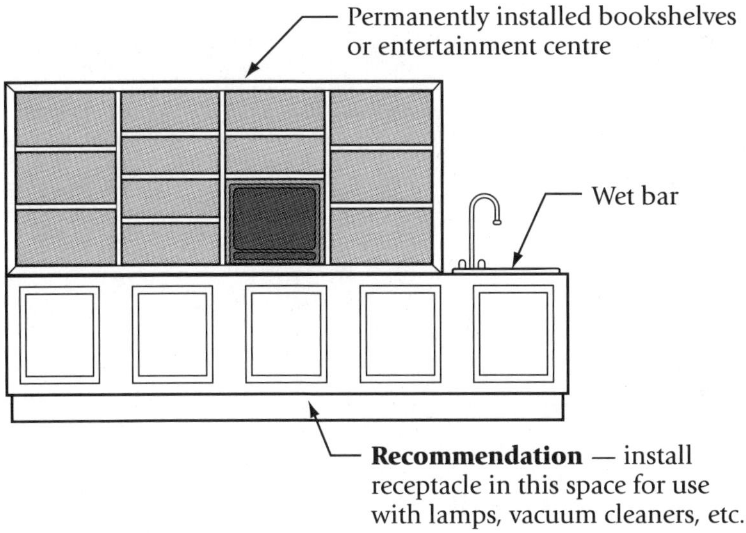


**(1)** Wall length behind these bookshelves or entertainment centre is unusable.

**(2)** "Permanently installed" means connected to the wall by devices that require a tool to be used to move the item.

**Figure 20 — Detail 4**

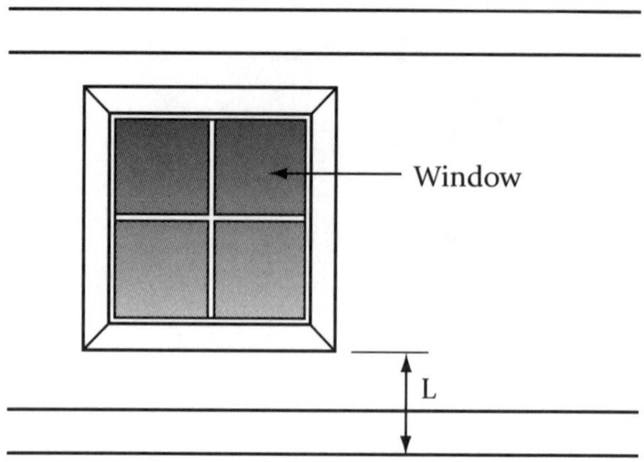


L is more than 300 mm (12 in) — usable wall space

**Figure 20 — Detail 5**

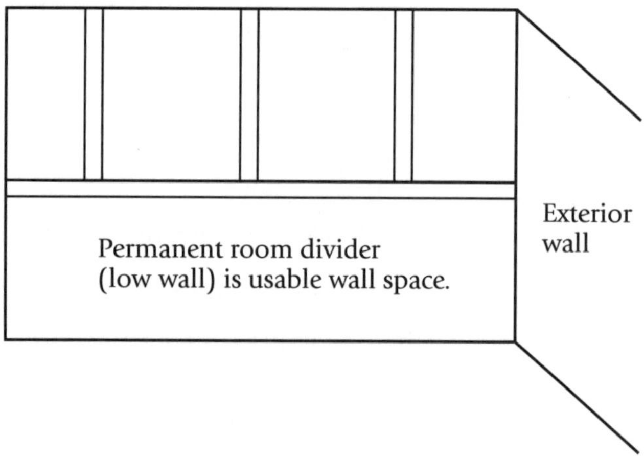


**Figure 20 — Detail 6**

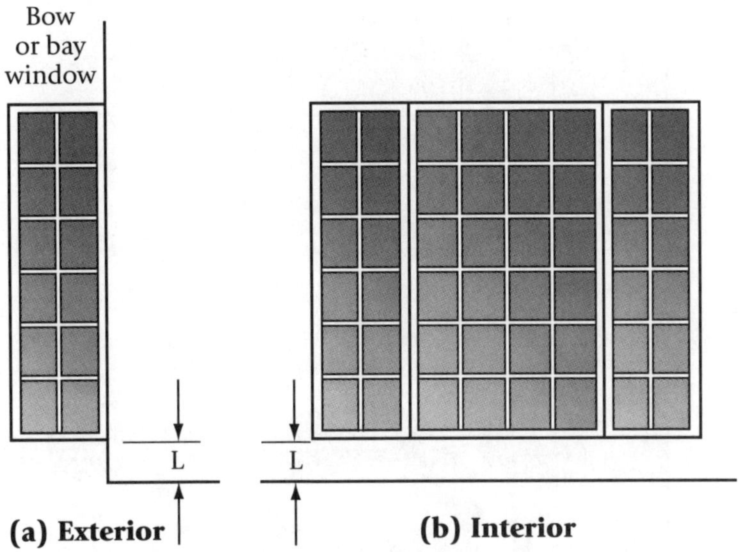


L is 300 mm (12 in) or less — the space below the window is unusable wall space

Figure 20 — Detail 7

The wall space behind the fireplace is unusable wall space.

**Figure 20 — Detail 8**

## What is the minimum number of receptacles required for a finished wall?

The Code requires that no point along the floor line of any usable wall space be more than 1.8 m (6 ft) horizontally from a receptacle in that wall space or an adjoining space, with the distance measured along the floor line of the wall spaces involved. This requirement is sometimes called the "six foot, twelve foot, six foot" rule. It means that the first receptacle in a finished wall must be located no more than 1.8 m (6 ft) from the start of the finished wall, the next receptacle must be a maximum of 3.6 m (12 ft) from the previous receptacle, and the last receptacle must be a maximum of 1.8 m (6 ft) from the end of the finished wall. All measurements are taken along the perimeter of the finished wall, at the floor line.

The rule is applied as follows:

**Step 1** Identify any unusable wall spaces around the perimeter of the room or area. See "What is usable wall space?" on page 54.

**Step 2** Measure the first 1.8 m (6 ft) from the start of a finished wall along the perimeter of the usable wall space, at the floor line, to determine where the first receptacle should be. It can be anywhere within the first 1.8 m.

**Step 3** From the location of the first receptacle, measure 3.6 m (12 ft) along the wall perimeter, at the floor line. The second receptacle can be located anywhere within the 3.6 m distance from the first receptacle.

**Step 4** Repeat Step 3 until the end of the perimeter of the finished wall has been reached.

**Step 5** The last receptacle in the series must be within 1.8 m (6 ft) of the end of the finished wall.

See Figures 21 to 23.

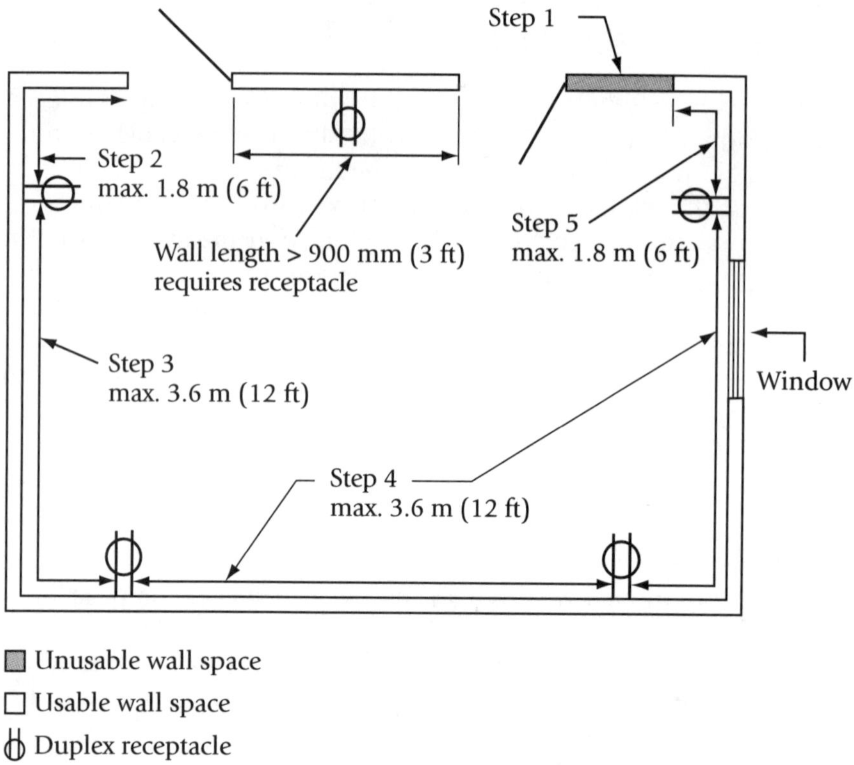


**Figure 21**
**Finished walls — Bedroom**

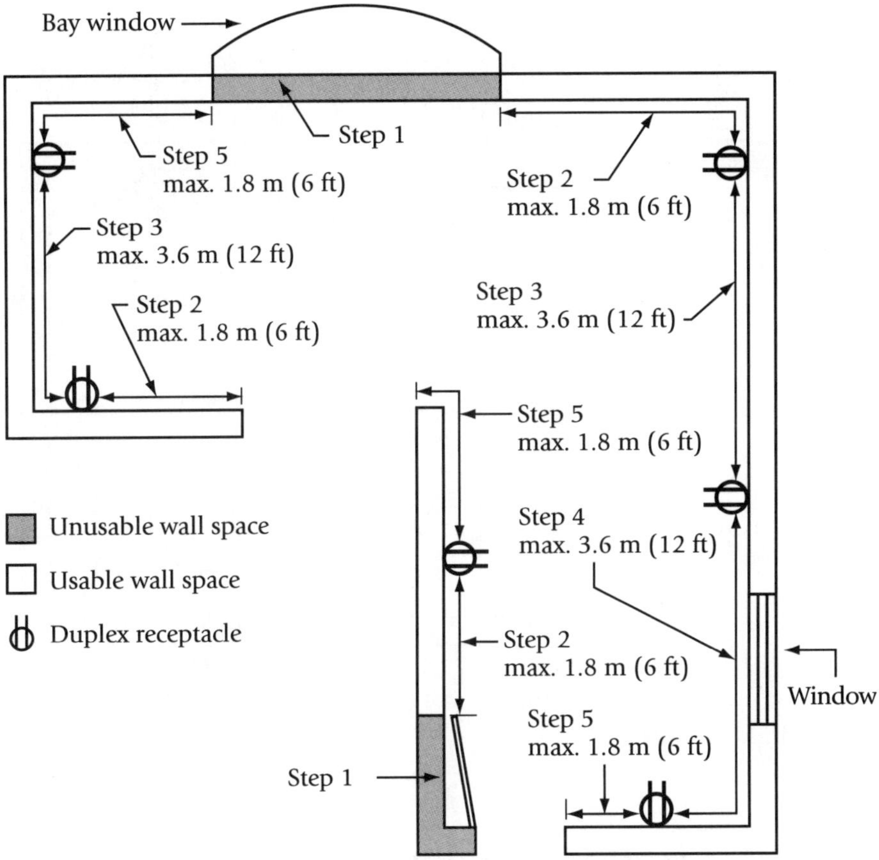


**Figure 22**
**Finished walls — Living/dining room**

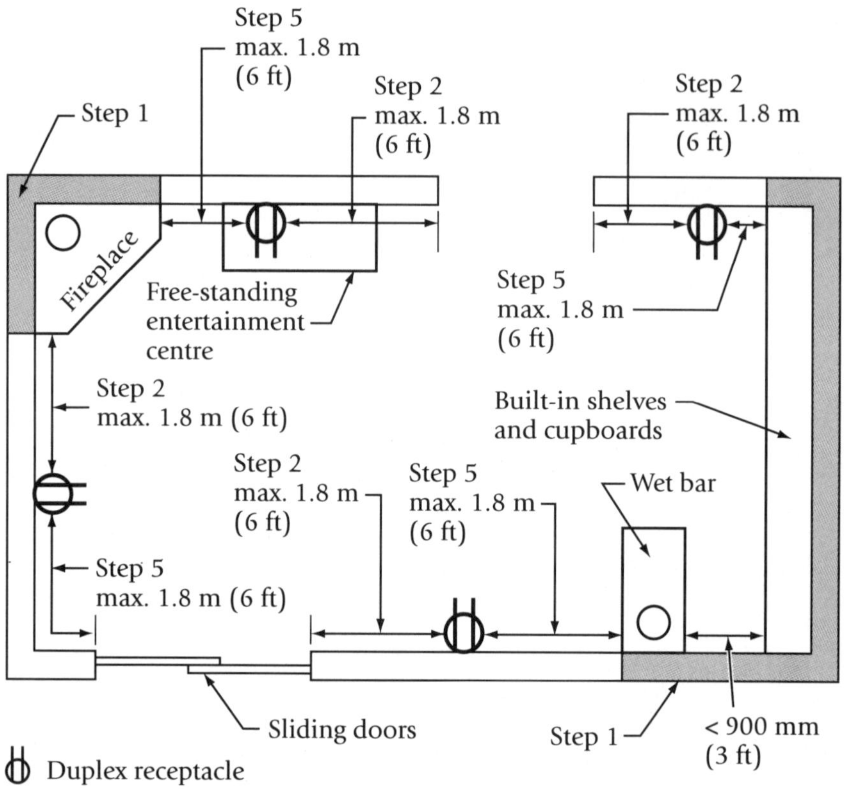


**Figure 23**
**Finished walls — Recreation room**

## What are the receptacle and circuit requirements for finished walls?

For finished walls other than those in sleeping rooms and areas (see "What are the receptacle and circuit requirements for finished walls in sleeping rooms and areas?" below), general-purpose receptacles are required. Each receptacle is one of the 12 outlets permitted on a 15 A, 120 V, two-wire general-purpose circuit.

A maximum of 12 receptacles and lighting outlets may be installed on a general-purpose circuit protected by a 15 A overcurrent device (see "What is a general-purpose circuit?" on page 45). If only receptacles are installed on the circuit, as many as twelve 15 A, 120 V duplex receptacles can be installed in a dwelling's finished walls, protected by a single 15 A fuse or circuit breaker. However, it is a good idea to install fewer receptacles on a general-purpose circuit to allow for future expansion; it might become necessary to install another receptacle, should decorating requirements change or the function of an area be altered (for instance, converting a room into a home office). The load on the existing outlets might also be increased by the use of new electrical equipment — appliances, entertainment centres, computers, scanners, printers, etc. Consequently, a maximum of ten outlets is recommended, to make it easier to handle future requirements.

## What are the receptacle and circuit requirements for finished walls in sleeping rooms and areas?

The definition of finished wall and the identification of usable wall space are the same for sleeping rooms and areas as for other rooms in a dwelling. As well, the requirements for the kind and number of receptacles are the same, and the recommended maximum of ten receptacles and/or lighting outlets also applies. While sleeping rooms and areas are also supplied by general-purpose circuits, there is an important difference in the kind of overcurrent protection provided in the panel for these circuits: they must be provided with an arc fault circuit interrupter (AFCI).

Every circuit breaker or fuse used for branch circuit protection in dwellings is designed to protect circuits from damage due to excessive heating or short-circuit conditions, by opening automatically. The response time of circuit breakers and fuses is determined by the magnitude and duration of an overcurrent condition. A very high load, even if very short in duration, will trip the overcurrent device; so will a lower overload that goes on for a considerable time. This mechanical response is enough to protect a circuit so long as the integrity of the circuit conductors is maintained. However, if this integrity is compromised by the deterioration of insulation or the cracking or breaking of conductors, electrical arcing may occur. The temperatures associated with these arcs may exceed 6000 °C.

Statistical data have shown that in dwellings, risks from fires ignited by persistent electrical arcing are highest in sleeping areas. In uninhabited bedrooms, sleeping lofts, and so on, a fire can break out and remain unnoticed until it is beyond control. The risk is even greater when people are sleeping in these areas and can be overcome by smoke or the rapid advance of a fire.Because of these heightened risks, the Code requires that an arc fault circuit interrupter be installed on any circuit that supplies a sleeping area (see Figure 16 on page 33). An AFCI recognizes the current or voltage characteristics (or a combination thereof) associated with arcing faults and opens the circuit when an arc fault is detected. In this way, an AFCI can mitigate the risks posed by smoke or fire in sleeping facilities.

An AFCI is not intended to detect hot or glowing connections (caused, for example, by loose or corroded connections) that cause damage to the insulation by heating it above its temperature rating (see "What is a panel?" on page 29). If the circuit in which an AFCI is installed also requires other kinds of protection, such as ground fault circuit interruption, the AFCI must also comply with the applicable requirements of CSA Standards for the devices that perform those required functions.

The Code requires that only receptacles in sleeping rooms and facilities must be protected by an AFCI. Other types of outlets, such as lighting outlets, may also be connected to a circuit protected by an AFCI. However, care should be taken when connecting outlets other than receptacles to such a circuit. If an arc fault occurs, the AFCI will shut off all the outlets connected to it, and if these outlets include lighting, an area could be left in total darkness, creating a safety hazard.

## What are the requirements for receptacles in porches and balconies?

Many dwellings have attached balconies or porches, which may or may not be enclosed from the outside elements. It is common for people to use appliances, entertainment devices, lights, decorations, etc., on these porches and balconies.

Because of safety concerns over the use of extension cords and inadequate homemade wiring solutions, the Code requires that at least one duplex receptacle be installed in these locations. However, there are no strict requirements for the number of receptacles or for their location along the perimeter of the wall.

Each receptacle is one of the 12 outlets allowed on a 15 A, 120 V, two-wire general-purpose circuit that is protected by a standard fuse or circuit breaker. If the duplex receptacle is exposed to the weather, it must be provided with weatherproof cover plates. Because a shock hazard can occur when people standing on wet concrete or earth use energized electrical equipment, the Code also requires that the receptacles be protected by a ground fault circuit interrupter unless they are installed at least 2.5 m (8 ft) above ground level (for example, on an elevated porch or on a second-floor balcony; see Figure 42 on page 102).

# What are the requirements for receptacles in kitchens?

## What are the special factors in kitchens?

In dwellings, the kitchen is usually the area that requires the greatest number of branch circuits, and the greatest variety of voltage and current ratings for these branch circuits, because of the large number of appliances that are typically installed and used. These range from low- and high-powered portable devices to large appliances such as ranges, refrigerators, and dishwashers. Kitchens are often redone or renovated, with the addition of new appliances and upgraded equipment, so kitchen wiring must also take into account possible future demands.

While most kitchens have the same basic set of appliances, the design of appliances and the layout of equipment vary enormously, depending on an owner's personal requirements and preferences. Before beginning the layout of a kitchen's electrical installation in a new construction or in extensive renovations, it is crucial to have the plans for the kitchen layout finalized by the builder or owner. Knowing the locations and sizes of large appliances and the configuration of the cabinetry makes it possible to determine where various outlets, mandated by the Code and/or required by the owner, must be installed. Without the final layout, receptacles may be located incorrectly or mounted at the wrong height, resulting in costly alterations.

Kitchens are often very humid. When installing receptacles, it is important to make sure that the vapour, air, and thermal barriers required on outside walls and ceilings are behind the outlet boxes. Otherwise, in cold weather, cold air could come through the outlet box. As well as being drafty, this could cause frost to form on the device connected to the outlet box, which could in turn cause rapid deterioration of the device and connections, as well as presenting a possible shock hazard.

## What are the requirements for dining and eating areas in the kitchen?

### General requirements

Many kitchens are large enough to contain an eating or dining area in the form of an island, table and chairs, or built-in booth. To allow for the use of kitchen appliances, such as coffee pots, warming dishes, and toasters, the Code requires at least one duplex receptacle in such areas.

In some dwellings with an open style of architecture, it is sometimes difficult to determine whether or not a dining or eating area is separate from the kitchen. If the area is part of the kitchen, the requirements discussed in this section apply. If the area is considered a separate dining room or area — that is, if it is separated from the kitchen by a wall, which may or may not have an opening (with or without a door) that communicates with the kitchen — then the general receptacle requirements for finished walls apply. When in doubt, consult the local inspection department or simply install the required separate duplex receptacle, along with the receptacles supplied in accordance with the rules for finished walls.

### Circuit requirements

The Code requires that a separate branch circuit be used for the duplex receptacles in the dining area of a kitchen. More than one duplex receptacle can be connected to the branch circuit, provided that all the receptacles are in the dining area of the kitchen. There is one exception: a receptacle installed behind a free-standing gas range, which is only for the use of the gas range, may also be connected to the dining area receptacle circuit. (See "What are the requirements for a free-standing gas range?" on page 87.)

## What are the requirements for kitchen counter work surfaces?

### General requirements

Kitchen counters and work surfaces are areas where a number of portable appliances may be connected to receptacles and used at the same time. Many of these kitchen appliances have ratings between 1000 and 1500 W at 120 V. Depending on their ratings, if two or three appliances such as toaster ovens, blenders, electric fryers, and coffee makers were all on the same branch circuit, the overcurrent device would do what it is designed to do and trip because of the overload.

Anticipating owners' requirements for cooking and preparing food, the Code requires the following:

- The minimum number of receptacles available at the kitchen counter work surface must be directly proportional to the length of the usable space along the kitchen counter work surface in contact with the wall. The longer the usable space on the counter, the more likely it is that more appliances will be used.
- The number of receptacles allowed on a branch circuit must be reduced to limit the number of appliances that can be used at the same time on the same circuit. This reduces the likelihood of the overcurrent device tripping.

### Usable space on a kitchen counter work surface

The usable space on a kitchen counter work surface is determined by the perimeter of the surface that is in contact with the wall. This means that there is no usable space on island counters or on counters that extend perpendicular to a wall (peninsular counter work spaces). "What are the requirements for kitchen islands and peninsular counters?" on page 82 describes the requirements for receptacles on a kitchen island or a peninsular work surface. Areas behind kitchen sinks are not considered to be usable space, and receptacles must not be installed in such locations. Receptacles in this location would be subject to splashing water, and hazards could be created if attachment cords came into contact with water or if attachment plugs were inserted or removed by people with wet hands.

The usable counter space also excludes

- built-in equipment (for example, enclosures for microwave ovens); and
- any isolated work surface less than 300 mm (1 ft) wide, measured along the wall line.

## Determining the minimum number of receptacles

The Code requires that no point measured along the wall behind usable counter space be more than 900 mm (3 ft) from a receptacle. The first receptacle in the wall behind the counter must be located no more than 900 mm (3 ft) from the beginning of the counter; each following receptacle must be a maximum of 1800 mm (6 ft) from the previous receptacle; and the last receptacle must be a maximum of 900 mm (3 ft) from the end of the work surface. All measurements are taken along the perimeter of the counter work surface at the wall line.

The rule is applied as follows:

**Step 1** Determine any unusable wall space along the wall line of the kitchen counter work surface.

**Step 2** Measure the first 900 mm (3 ft) from the beginning of the counter work surface, at the wall line, along the perimeter of the counter. The first receptacle can be located anywhere within this 900 mm length.

**Step 3** From the first receptacle, measure 1800 mm (6 ft) along the counter, at the wall line. The next receptacle can be located anywhere within this 1800 mm distance from the first receptacle.

**Step 4** Repeat Step 3 until the end of the perimeter of the counter work surface is reached. (Note, however, that few kitchens have an uninterrupted usable wall space in excess of 3.6 m (12 ft). In most cases, proceed directly to Step 5.)

**Step 5** The last receptacle located in accordance with Step 3 or 4 must be within 900 mm (3 ft) of the end of the counter.

See Figures 24 to 26.

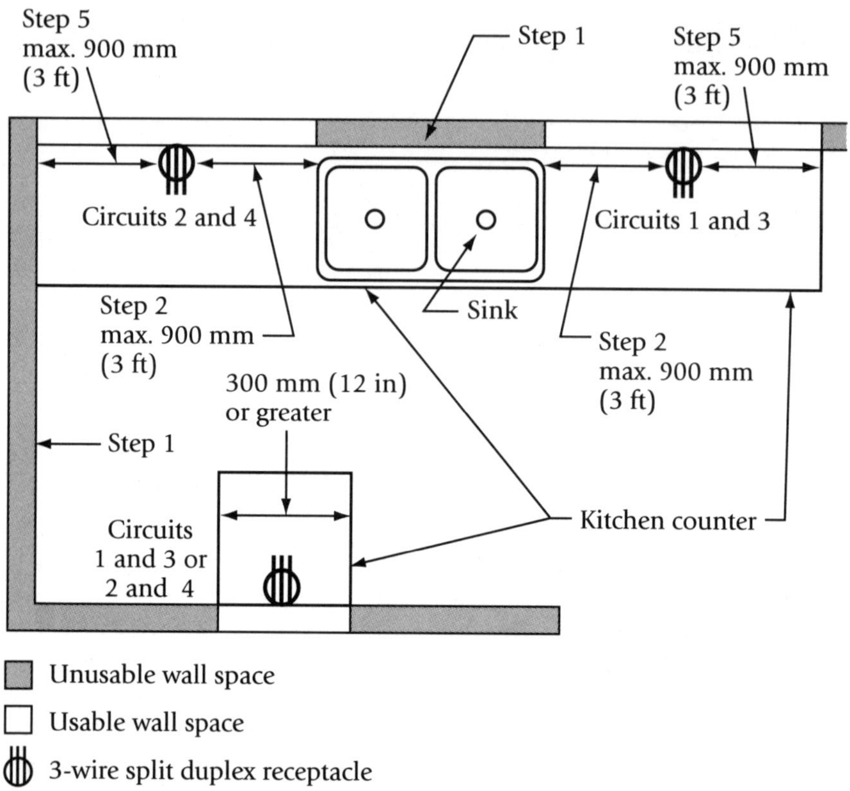


**Figure 24**
**Kitchen receptacles (1)**

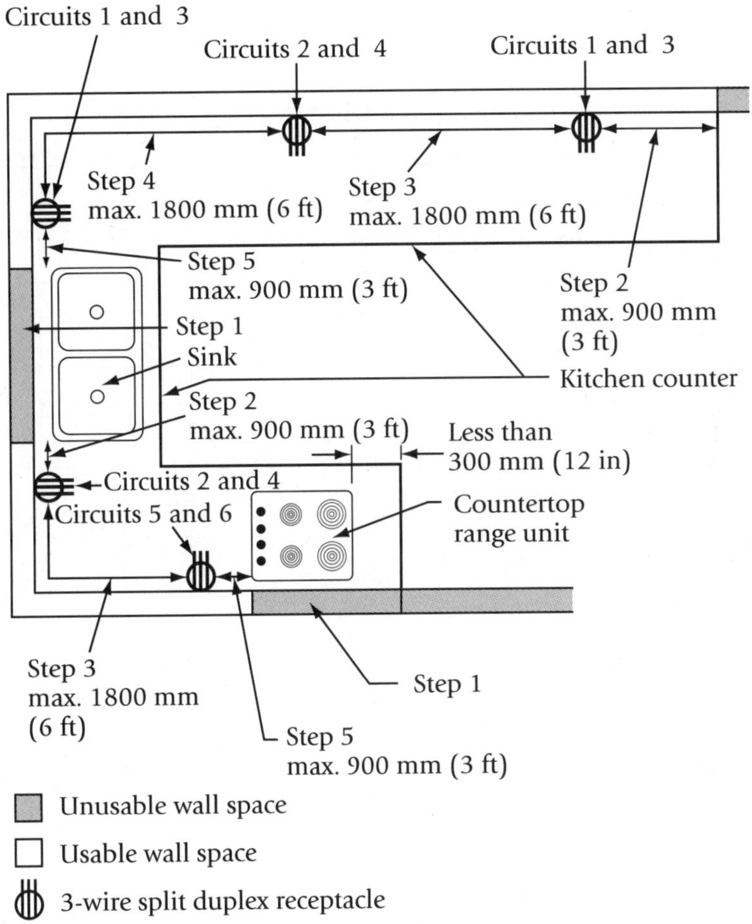


**Figure 25**
**Kitchen receptacles (2)**

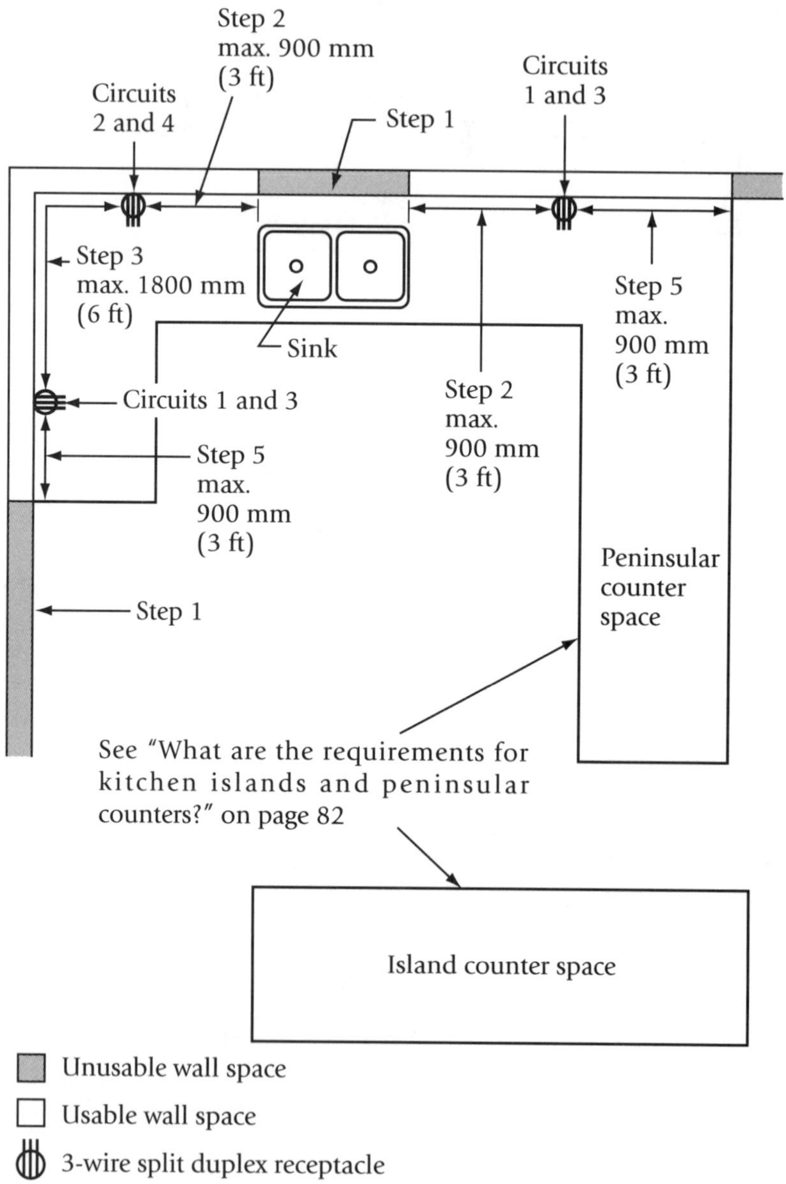


**Figure 26**
**Kitchen receptacles (3)**

## Receptacle and circuit requirements

Two types of receptacle are approved for use at kitchen counter work surfaces:

- a 5-15R split-duplex receptacle supplied by a 15 A multi-wire (or three-wire) branch circuit; and
- a 5-20RA duplex receptacle (also called a 20 A T-slot receptacle) supplied by a 20 A two-wire branch circuit.

No matter which kind of receptacle is used, the Code requires that

- there be no more than two duplex receptacles on any one branch circuit;
- there be no other outlets or receptacles connected to the branch circuit; and
- adjacent duplex receptacles not be connected to the same branch circuit.

One exception is made to these general requirements. In a kitchen designed for wheelchair access, one split-duplex receptacle that is installed in the front of a kitchen counter may be connected to a branch circuit that supplies two split-duplex receptacles located on the wall behind the counter work surface (see "What are the requirements for kitchen counter work surfaces designed for wheelchair access?" on page 80).

## Split-duplex receptacle

The split-duplex receptacle used on a multi-wire branch circuit is a standard duplex receptacle, which is manufactured with a break-off tab between the two line terminals (the brass-coloured screws) and the two neutral terminals (the silver-coloured screws). On a split-duplex receptacle, the tab between the line terminals is removed, allowing each half of the receptacle to be on a different 15 A, 120 V branch circuit (see Figure 27).

**Note:** *Only the tab on the line side of the receptacle should be broken off. The line side is usually the side that has brass-coloured terminal screws and a shorter, rectangular slot for a receptacle blade.*

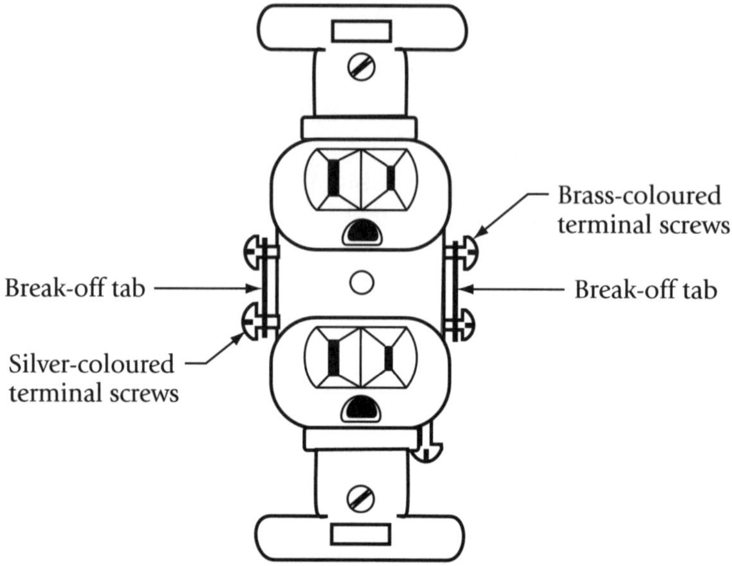


**Figure 27**
**Duplex receptacle 5-15R**

The multi-wire branch circuit that supplies the split-duplex receptacle consists of three conductors in a cable: two insulated line conductors, which are normally red and black, and a neutral insulated conductor, which is white. There is also a bonding conductor in the cable, which is usually bare and is not counted in the conductors in the cable. There is a line voltage of 240 V between the two line conductors and of 120 V between the neutral conductor and each line conductor (see Figure 28). The multi-wire circuit is protected by two standard 15 A circuit breakers or fuses in the panel, usually installed as a double-pole circuit breaker, two single-pole breakers with their handles tied together, or a 240 V fuse block.

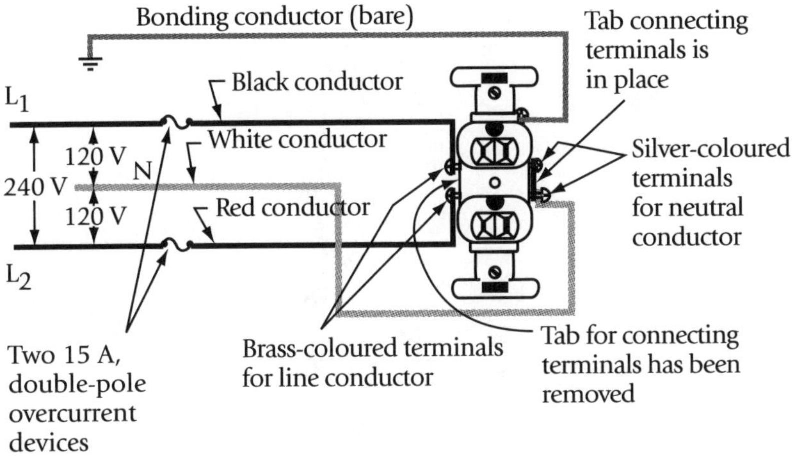


Figure 28
**Multi-wire branch circuit to a 5-15R split-duplex receptacle**

## T-slot receptacles

A 20 A T-slot (or 5-20RA duplex) receptacle gets its name from the neutral slot, which has the shape of the letter "T" on its side when the receptacle is in the vertical position (see Figure 29). The receptacle accepts both 15 A, 120 V and 20 A, 120 V male plugs for connecting portable appliances. It is supplied by a 20 A, 120 V, two-wire branch circuit that consists of an insulated line conductor, usually black, and a white insulated neutral conductor. There is a line voltage of 120 V between the neutral conductor and the line conductor.

The T-slot receptacle accepts two plugs and supplies both plugs with only one 20 A, 120 V branch circuit, compared to the split-duplex receptacle, which supplies two 15 A, 120 V branch circuits, one for each plug.

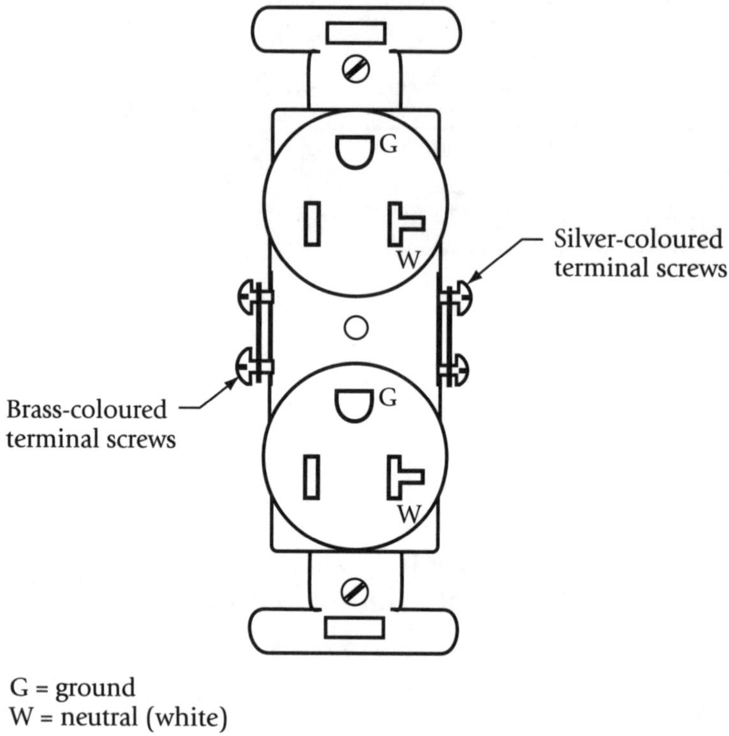


Figure 29
**5-20RA duplex or 20 A T-slot receptacle**

## Split-duplex or T-slot — Which is better?

There is no simple answer to this question. The 15 A split-duplex receptacle has been the Code-approved standard in Canada for many years, and the 20 A T-slot has been the standard in the United States, where it is approved by the *National Electrical Code*. In 1998, in an effort to harmonize electrical requirements in Canada and the U.S., Canada adopted the T-slot as a permissible alternative to the split-duplex receptacle.

Choosing between the two depends on

- cost: the 14/3 cable used to wire split-duplex receptacles is more expensive than the 12/2 cable used for T-slot receptacles;
- convenience for the installer: it may be easier to use just one size of cable, a #14, rather than two sizes of cable, a #12 and a #14, which would be required for a combination of 15 A receptacles and 20 A T-slot receptacles; and
- intended use: high appliance loading at the receptacle might require the heavier-duty service of a multi-wire circuit and a split-duplex receptacle.

Another important consideration is whether or not the receptacle needs to be protected by a ground fault circuit interrupter (GFCI). The Code requires that any receptacle within 3 m (10 ft) of a sink, bathtub, or shower stall be protected by a Class A GFCI, and in some parts of Canada, ground fault protection is also required for any receptacle located on a kitchen counter work surface that contains a sink (check with local inspection authorities). T-slot duplex receptacles with built-in GFCIs cost significantly less than the two-pole GFCI circuit breakers that have to be installed in the panel to protect 15 A split-duplex receptacles.

The following comparison chart will help you to select the better option:

| Requirement | Split-duplex receptacle | T-slot receptacle |
|---|---|---|
| Minimum size of conductor | #14 AWG | #12 AWG |
| Maximum rating of overcurrent device | 15 A, double-pole | 20 A, single-pole |
| Minimum number of circuit conductors | 3 (2 line, 1 neutral) | 2 (1 line, 1 neutral) |
| Maximum number of receptacles on circuit for kitchen counter work surfaces | 2 (one additional receptacle allowed in the front of the counter when installed for wheelchair accessibility) | 2 (one additional receptacle allowed in the front of the counter when installed for wheelchair accessibility) |
| Number of circuits at receptacle | Two 15 A circuits | One 20 A circuit |

## What are the requirements for kitchen counter work surfaces designed for wheelchair access?

The Code recognizes that in a kitchen designed for persons with disabilities, it is convenient to install receptacles on the front of a counter work surface. These receptacles are provided in addition to those required for the usable counter space in accordance with the Code and are not intended to replace the receptacles that must be located on the wall behind the work surface (see Figure 30). Although the Code generally specifies a maximum of two kitchen counter split-duplex receptacles on a single multi-wire circuit, a third receptacle can be connected in such installations because it is considered unlikely that all three will be used at the same time.

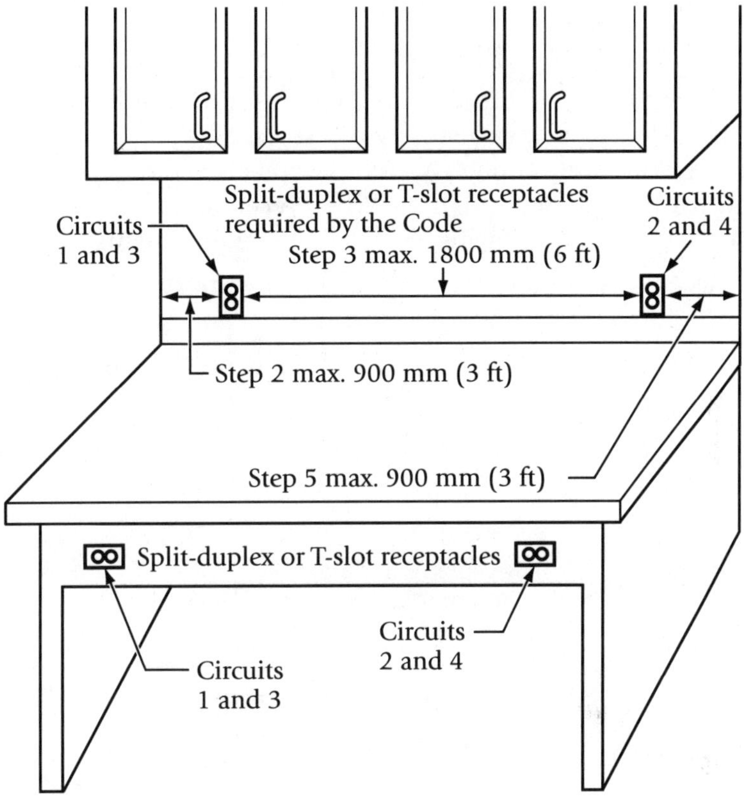


**Note:** *See "Determining the minimum number of receptacles" on page 71.*

Figure 30
**Kitchen work surface designed for persons with disabilities — Supplementary split-duplex or T-slot receptacles**

## What are the requirements for kitchen islands and peninsular counters?

At the present time, the Code neither requires nor prohibits receptacles on kitchen islands or peninsular work surfaces. People often want to have at least one split-duplex or T-slot receptacle on these work surfaces to allow for the use of appliances. When receptacles are installed, they must conform to the following general Code requirements:

- Receptacles in kitchen or dining area work surfaces must not be mounted facing up (see Figure 31).
- The receptacles must not be installed on the side of the work surface directly in front of any sink.

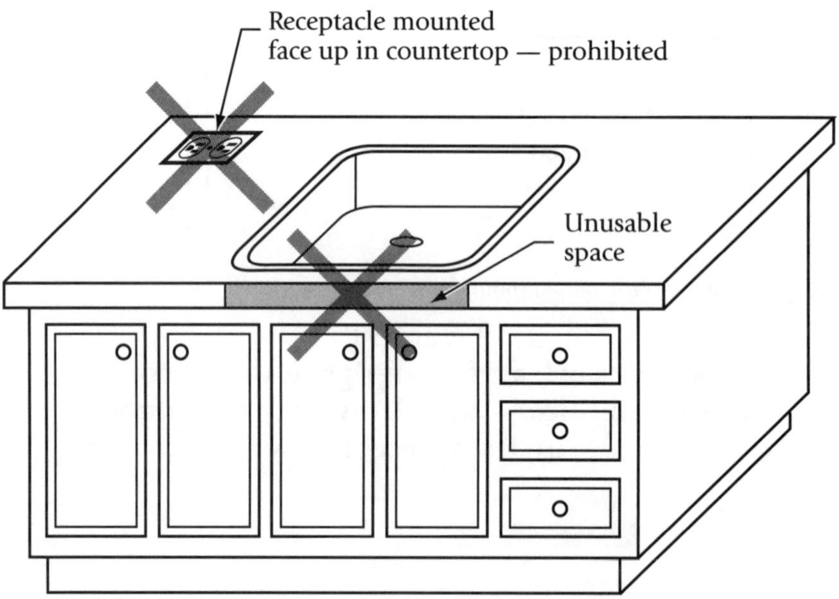


**Figure 31**
**Prohibited surface mounting**

Consequently, receptacles on islands or peninsular counters are installed either in an approved, enclosed floor outlet box ("floor monument") on the top surface of the counter or on the side or front of the counter not directly in front of any sink (see Figure 32). When cable is run inside a cabinet to feed receptacles either on a floor monument on the top surface or on the side or front of the counter, it must be protected from damage. One option is to use armoured cable; another is to enclose nonmetallic sheathed cable in flexible conduit, electrical metallic tubing, or PVC conduit.

Receptacles installed on islands and peninsular work surfaces pose a potential safety hazard. In many cases, appliance cords plugged into these receptacles will be long enough to hang over the edge of the work surface, making it easy to pull the appliance off the work surface by accidentally catching or pulling the appliance cord.

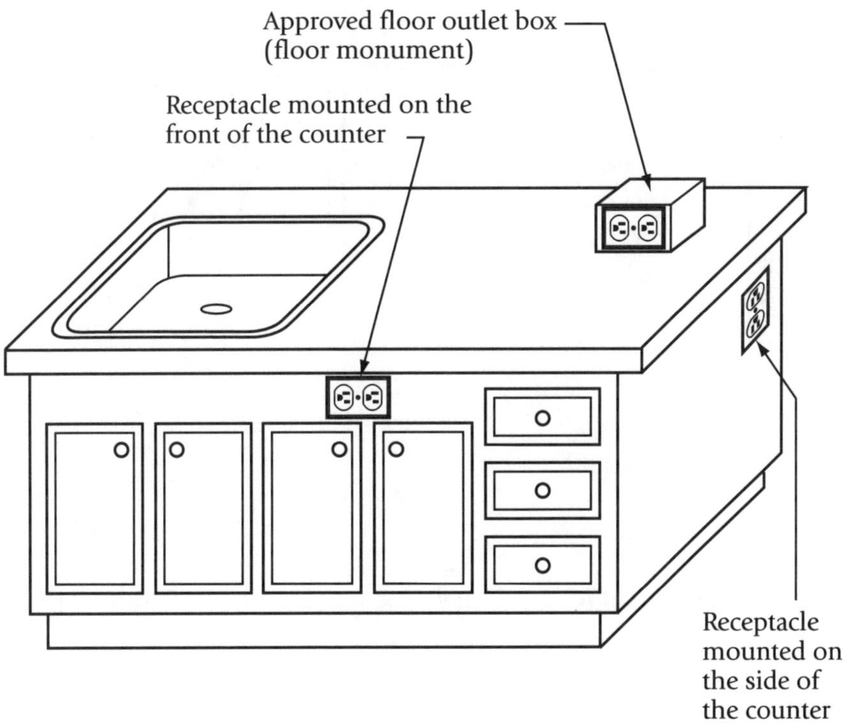


Figure 32
**Permitted locations for receptacles on kitchen islands and peninsular counters**

## What are the requirements for a refrigerator?

### Receptacle requirements

In a dwelling, the Code requires that the kitchen have one receptacle that is used for the refrigerator only. The receptacle is installed behind the refrigerator, and it can be either a duplex or a single-socket receptacle (see Figure 33). Duplex receptacles are most often used because they are cheaper and readily available, even though only one appliance will be connected to the receptacle. The Code does not specify mounting heights, but the generally accepted trade practice is a wall mounting between 600 mm (2 ft) and 1200 mm (4 ft) above the finished floor.

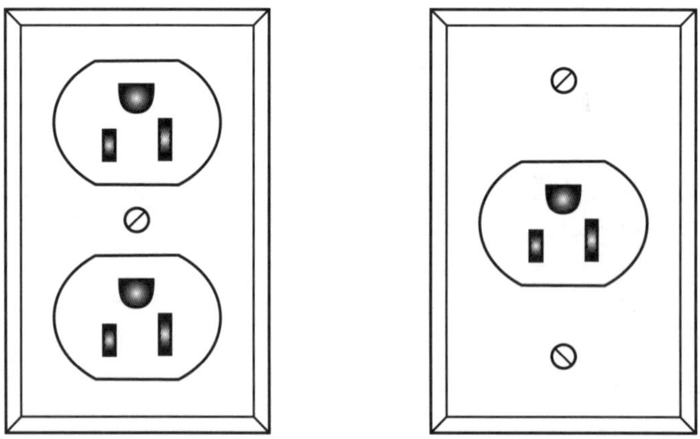

**Duplex receptacle**

**Single receptacle**

Figure 33
**Types of receptacles — Duplex and single for refrigerators**

### Circuit requirements

The Code requires that the refrigerator receptacle be supplied by a separate branch circuit that has no other outlets connected to it, with the permitted exception of a recessed clock outlet. Refrigerators draw a considerable amount of electrical power, but an electric clock uses only a very small amount, so the Code allows such a clock to be installed on the circuit with the refrigerator receptacle. The recessed clock outlet is a device box containing a recessed single receptacle. When the plug of the clock is inserted and the clock is attached to the device box, usually by a hook, the clock is flush with the wall surface, and the whole outlet is hidden behind the clock.

## What are the requirements for a free-standing electric range?

### Receptacle requirements

If the free-standing electric range has a demand rating of 50 A or less (approximately 16 kW or less), the Code requires the installation of a 14-50R receptacle that

- is rated 125/250 V, 50 A;
- is mounted 130 mm (5 in) to the centre of the receptacle above the finished floor;
- is mounted in the centre of the wall space provided for the range; and
- has the ground U-slot oriented to either side.

See Figures 34 and 35.

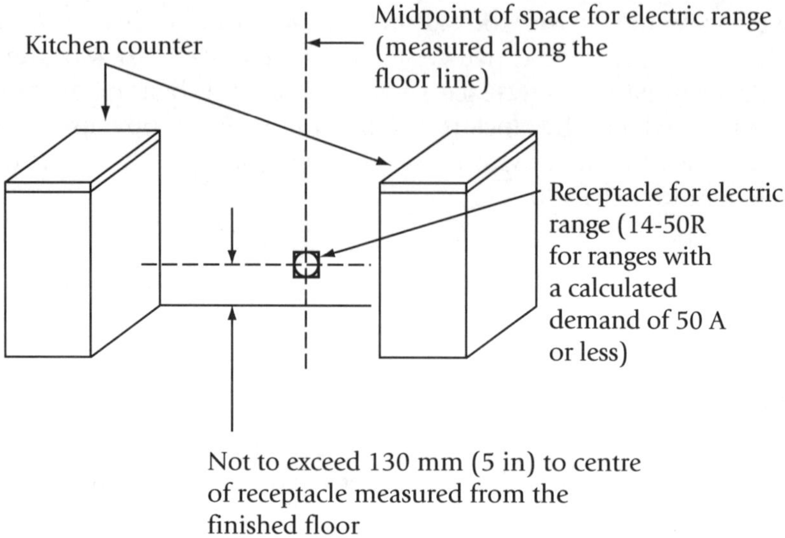


**Figure 34**
**Receptacle location for a free-standing electric range**

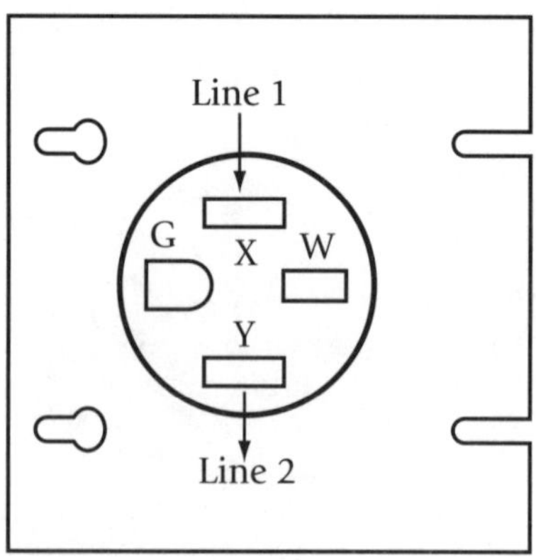


G = ground
W = neutral

**Figure 35**
**14-50R receptacle — Orientation**

The location requirements ensure that the receptacle will be accommodated by the standard recess left at the back of a range so that it can be installed right up against the wall. The male plug for a range is approximately 38 mm (1½ in) thick, and if the receptacle were incorrectly located, it might prevent a proper installation. These two requirements also make it possible to keep the large range cord that connects the range to the receptacle well out of the way, in the recess behind the range, where it is less likely to be damaged.

### Circuit requirements

The Code requires that a free-standing electric range be supplied by a dedicated separate branch circuit sized according to the rating of the range (see "How are branch circuits sized?" on page 50). For a range that does not exceed 16 kW at 120/240 V, a 40 A, 240 V, double-pole overcurrent device, as a maximum, is used to protect a branch circuit that consists of a three-conductor #8 copper cable. The cable type varies depending on where the cable is being used (combustible or noncombustible construction, concealed or exposed, and protected from or subject to mechanical injury). In wood frame construction, which uses wooden studs and joists and is considered combustible construction, the cable specified for use is nonmetallic 8-3 NMD90 FT1 copper cable (sometimes referred to as stove cable).

## What are the requirements for a free-standing gas range?

### Receptacle requirements

A free-standing gas range must be provided with one receptacle located behind the gas range. The Code requires the receptacle to be

- a 5-15R or a 5-20RA receptacle, either duplex or single;
- mounted 130 mm (5 in) to the centre of the receptacle above the finished floor; and
- mounted in the centre of the wall space provided for the range.

See Figure 36.

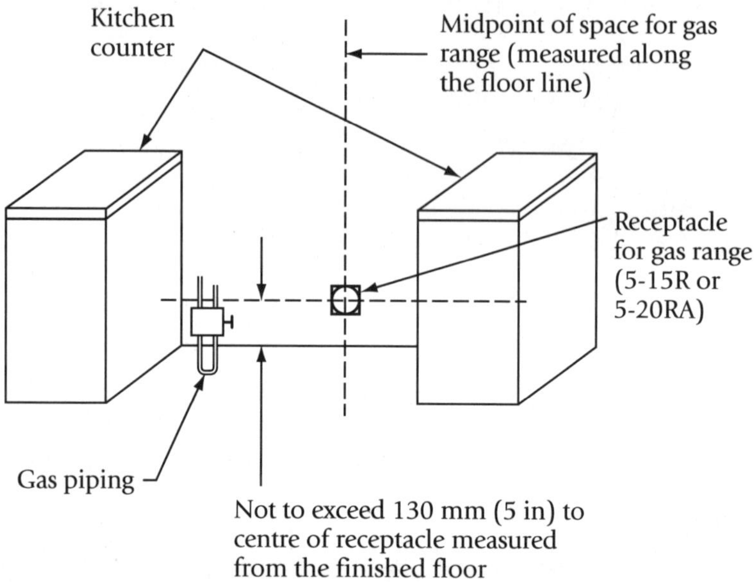


**Figure 36**
**Receptacle location for a free-standing gas range**

## Circuit requirements

The branch circuit that supplies the receptacle intended for use by a free-standing gas range may be

- connected to the same branch circuit used for the receptacle in the kitchen's dining or eating area (see "What are the requirements for dining and eating areas in the kitchen?" on page 69); or
- a general-use receptacle connected to a general-use 15 A, 120 V branch circuit (this might be the kitchen lighting circuit or the circuit to which the range's hood fan is connected).

## What are the requirements for a microwave oven?

Microwave ovens are cord-connected portable appliances and require a receptacle. The Code does not provide any requirements for microwave oven receptacles — beyond those that apply generally to work surface appliance receptacles — unless the microwave oven is to be installed in a cupboard, wall cabinet, or other enclosure.

In general, receptacles should not be placed in cupboards, cabinets, or similar enclosures in order to prevent the cord of an appliance from being damaged by someone trying to close a cupboard door on it, or to prevent a cooking or heating appliance from being plugged in and left on with the cupboard door closed, thus creating a severe fire hazard. However, the Code permits receptacles to be installed in a cupboard, cabinet, or a similar enclosure if the receptacle is energized only when the enclosure door is in the fully opened position. This usually requires a special enclosure in which the receptacle is an integral part of a factory-built enclosure. See Figure 37.

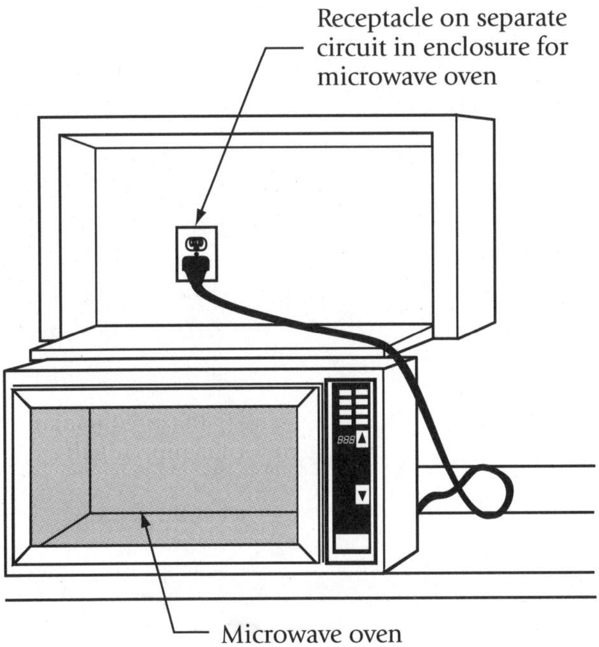


**Figure 37**
**Microwave cabinet**

# What are the requirements for receptacles in hallways and stairways?

## What special factors are involved?

Even though the walls in hallways and stairways usually qualify as finished walls, the Code requirements for the minimum number of receptacles do not apply. In dwellings, hallways and stairways serve mainly as corridors that allow occupants to move from room to room. Stairways have the added function of serving as a corridor between different floor levels. Neither hallways nor stairways are used by the occupants as cooking, eating, living, or sleeping areas. As a result, there is only a limited need for receptacles, which are used primarily for such things as cleaning equipment, communication devices (for example, telephones), and table lamps.

## What are the requirements for hallways and stairways in dwellings?

### Receptacle requirements

In dwellings, no receptacles need to be provided in stairways. However, the Code requires that receptacles be installed in hallways such that no point in the hallway is more than 4.5 m (15 ft) from a receptacle. Stairways are not included in calculating the length of a hallway. For example, where a second floor hallway is connected to the first floor hallway by a stairway, receptacles are provided according to the lengths of the two hallways treated as separate areas.

The number and location of the receptacles are determined as follows:

**Step 1** Measure 4.5 m (15 ft) from the corner of one end of the hallway. The first receptacle is installed anywhere within this 4.5 m length, usually in the most convenient wall location. (If the hallways are equipped with central vacuum outlets, it is a good idea to locate a receptacle near the vacuum outlet to supply the vacuum's electric-powered head.)

**Step 2** Measure 9 m (30 ft) from the first receptacle; the next receptacle can be located anywhere within this range. If the end of the hallway is less than 4.5 m (15 ft) from the first receptacle, no other receptacle is required.

The Code makes an exception to these general requirements when there is a receptacle on a finished wall inside an adjacent room that satisfies the 4.5 m (15 ft) rule, provided that the opening between the room and the hallway is not fitted with a door. For example, a hallway outside a living room might not require a receptacle because the entrances to living rooms (e.g., archways) are usually not fitted with a door.

### Circuit requirements

Hallway receptacles are general-purpose receptacles connected to a 15 A, 120 V, two-wire general-purpose circuit, which can have a maximum of 12 outlets. The hallway receptacles may be combined with other receptacle and lighting outlets on the circuit so long as the number of outlets does not exceed the maximum.

# What are the requirements for receptacles in utility and laundry rooms?

## What special factors are involved?

A room is generally defined as any part of a building enclosed by walls or partitions, a floor, and a ceiling. A utility room is any room or area designed primarily to segregate large equipment, such as furnaces, water heaters, and water softeners, from the common living spaces of a dwelling. A laundry room is a room or area intended mainly for the installation of large equipment such as clothes washers and dryers. Utility rooms and laundry rooms are essentially functional areas that provide useful services rather than serving as a living space. As a result, for the purpose of receptacle requirements, the Code does not consider these rooms to have finished walls, even if the walls are covered with drywall, wood, other types of panelling, or other surface finishes. The Code provides specific requirements for receptacles and supply circuits in laundry rooms and utility rooms.

## What are the requirements for utility rooms?

The Code requires that at least one duplex receptacle be installed in a utility room and that the receptacle be supplied by a separate branch circuit. Note, however, that not every area that contains equipment such as a furnace or water heater is considered a room. Consider the following questions:

- Is the equipment located in a room enclosed by walls, a floor, and a ceiling?
- Are the walls covered with drywall, wood, other types of panelling, or similar material?

A receptacle is required in an enclosed utility room with finished walls mainly to ensure that people servicing a furnace, water heater, water softener, or other equipment will not have to run extension cords through the doorway from another room (see Figure 38). The utility room receptacle usually supplies appliances with relatively low power requirements (work lights, vacuum cleaners, small power tools, etc.).

If the equipment is not enclosed, and the walls adjacent to the equipment are not covered with drywall, wood, or panelling — as in an unfinished basement, for example — the Code does not require a receptacle near the utility equipment. (One could easily be added at a later date without disturbing any wall materials.) The Code requires an unfinished basement to be provided with one duplex receptacle (see "What are the requirements for receptacles in unfinished basements?" on page 103), but it need not be supplied by a separate branch circuit.

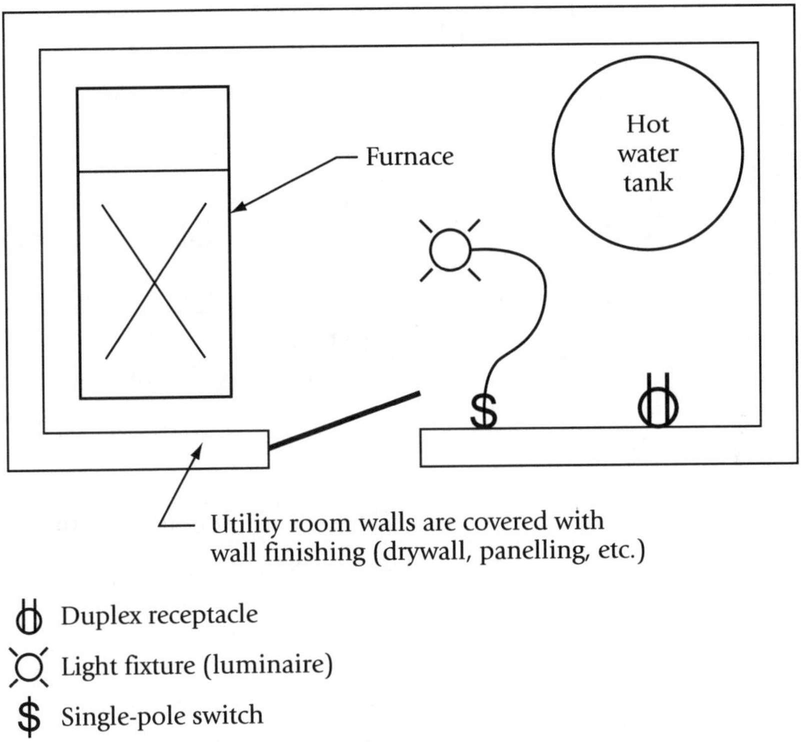


**Figure 38**
**Finished utility room**

## What are the requirements for laundry rooms and areas?

The Code refers to laundry rooms or areas because no distinction is made between an entirely enclosed room and a more open area in which large equipment such as clothes washers and dryers are installed and appliances such as irons are used. Wall finish is not taken into consideration in identifying laundry rooms or areas; circuit requirements are the principal concern because the appliances connected to these receptacles usually have relatively high energy ratings (1000 W or higher).

At least two duplex receptacles are required in laundry rooms or areas, one for a washing machine and one (or more) for appliances such as clothes irons and gas dryers. These receptacles must be supplied by a separate branch circuit that has no other outlets connected to it. Although laundry rooms and areas usually have exposed water supply piping and fixed laundry tubs, the Code does not consider these tubs to be sinks, and there is no requirement for duplex receptacles within 3 m (10 ft) of the laundry tub to be protected by a ground fault circuit interrupter, as there is in a bathroom (see Figure 39 below and "What are the circuit requirements?" on page 96). However, local requirements do vary, and in some parts of Canada a receptacle installed near a laundry tub may require GFCI protection. It is always a good practice to check with the local inspection department.

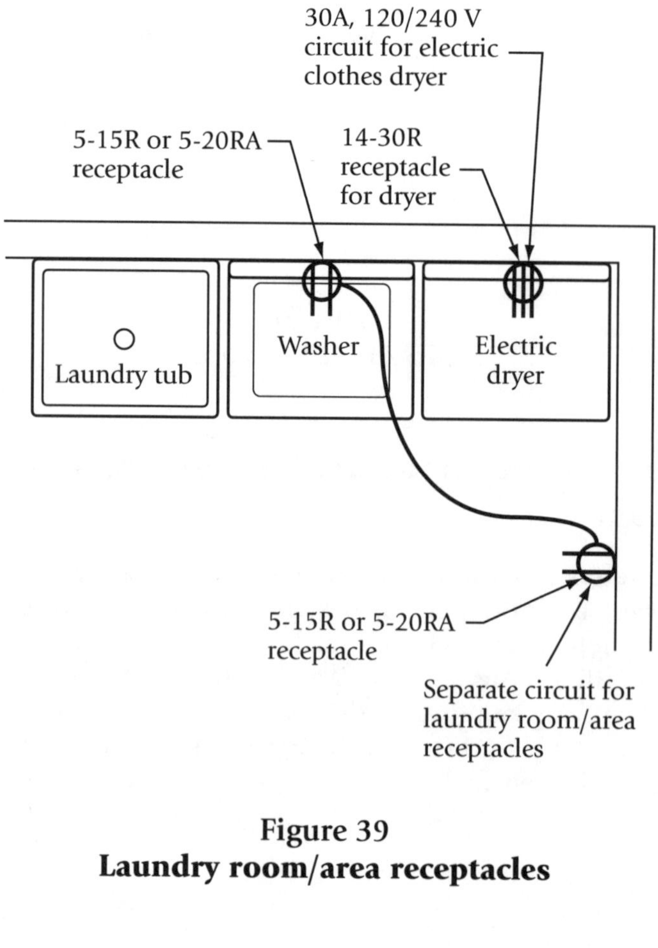


**Figure 39**
**Laundry room/area receptacles**

The Code does not give a mandatory mounting height for the clothes washer receptacle, but the standard practice is to locate the receptacle above the washer, usually 1.2 m (4 ft) from the floor. This makes it easy to unplug the washer for servicing, cleaning, and replacement.

In some dwellings, a combination washer-dryer unit is installed in a specially designed enclosure, similar to a closet. This is not considered a laundry room or area. These units are usually connected to the circuit by a cord set, and the receptacles provided for the units must conform to the Code rules for receptacles in enclosures. An additional receptacle is not required. The local inspection authority may also have requirements for this type of installation.

Laundry equipment may also be installed in bathrooms. The Code has some special requirements for this kind of laundry area; see "What are the requirements for bathrooms that contain laundry facilities?" on page 99.

## What are the requirements for receptacles in bathrooms?

### What are the receptacle requirements?

The Code requires at least one duplex receptacle to be located adjacent to the sink, no farther away than 1 m (3 ft). This receptacle is intended for appliances such as electric shavers, hair dryers, curling irons, and dental hygiene appliances that people use at the sink in a bathroom. (If the bathroom does not contain a sink, there is no mandatory requirement for receptacles.) The receptacle must be located above the level of the sink counter or above the level of the basin if it is a stand-alone type (a pedestal sink or a sink attached to the wall), and to one side, to allow a mirror to be installed over the sink.

In bathrooms that contain a bathtub or shower stall, there is a risk that a person in the tub or shower stall might use an appliance plugged into the receptacle near the sink, with consequent shock hazards. Therefore, the Code also requires that the duplex receptacle be located at least 1 m (3 ft) away from the bathtub or shower stall.

In some bathroom layouts, these two requirements conflict, and the Code allows an exception where it is simply not possible to place the receptacle at least 1 m (3 ft) away from the tub or shower. However, a minimum distance of 500 mm (20 in) is still required.

## What are the circuit requirements?

In general, the Code relies on insulation and grounding for protection against shock hazards. However, where the construction or function of an area could increase the risk of shock hazard, the Code requires additional protection. In the case of bathrooms, the Code requires that the sink receptacle, as well as any receptacle within 3 m (10 ft) of a sink, bathtub, or shower stall, be protected by a Class A ground fault circuit interrupter (GFCI). Ground fault protection for receptacles outside this limit is not mandatory, but the potential for shock hazard in these areas should still be considered. When in doubt, it is better — and usually no more difficult — to provide the ground fault protection.

A Class A GFCI trips the circuit connected to it when it detects a leakage current — the difference between the current that flows out in the circuit and the current flowing back — between 4 and 6 mA. The leakage current presents an electric shock hazard, and depending on its magnitude and duration, it can cause more or less harm to people and to the electrical installation.

GFCI protection is provided either by a GFCI circuit breaker in the panel or by a GFCI incorporated in the receptacle. A breaker with a built-in GFCI gives both overcurrent protection (against short circuit and excessive load) and ground fault protection to every outlet connected to the branch circuit. Where such a circuit breaker is used, the bathroom receptacles may be general-purpose duplex receptacles supplied by a general-purpose branch circuit, which can carry a maximum of 12 receptacles and outlets.

Two kinds of receptacles incorporate a GFCI: the stand-alone type, which provides ground fault protection only to devices and appliances that are plugged into it, and the feed-through type, which provides ground fault protection to devices and appliances that are plugged into it and to any other outlets or receptacles that are connected to its load terminals. Both receptacles are supplied by a general-purpose branch circuit, and both count toward the 12 outlet maximum that such a circuit can have. See Figure 40.

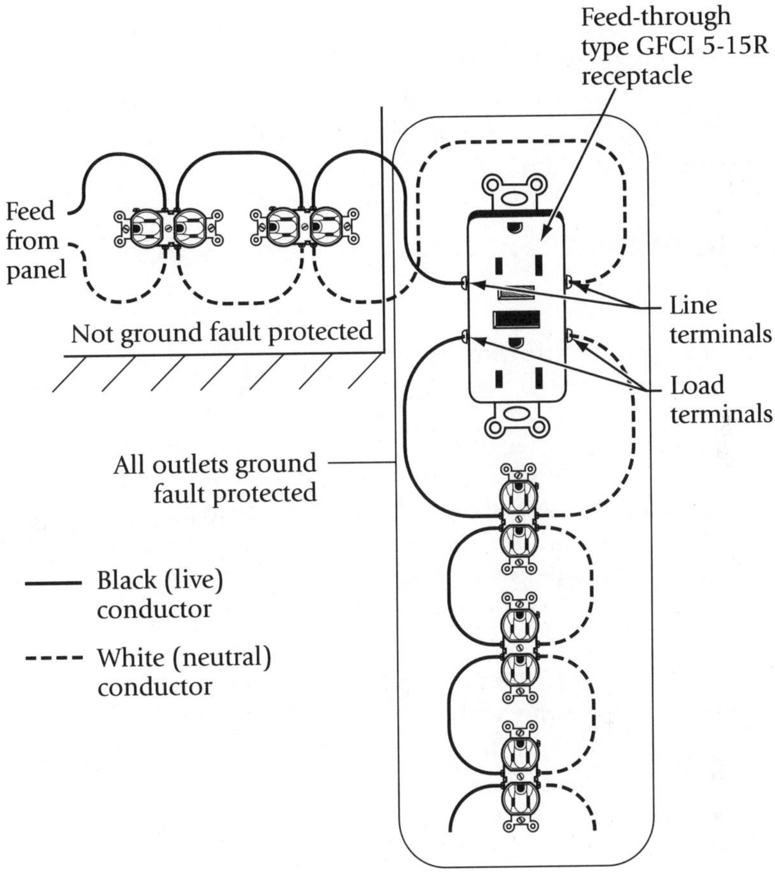


**Figure 40**
**Combination GFCI and general-purpose circuit**

Whether the circuit uses a GFCI breaker in the panel or GFCI receptacles, it is a good idea to use the circuit to supply only the receptacles in a bathroom, and to use another circuit to supply the lighting outlets. If the same circuit supplies both, the operation of the overcurrent protection would cut off the lighting as well, possibly creating a dangerous situation for someone in the room.

## How should circuits with GFCI protection be installed?

The ground fault circuit interrupter must be located where it is easily available for regular testing. Manufacturers generally have a monthly test as a condition of their warranty for GFCI devices.

As mentioned in "What are the general requirements for general-purpose circuits?" on page 46, it is always a good idea to connect fewer than the permitted maximum number of outlets to a general-purpose circuit. In bathrooms in particular, on a circuit with GFCI protection, reducing the number of outlets and receptacles to just five or six is advisable. This will help prevent nuisance tripping: the more outlets and receptacles connected to the interrupter, the greater the leakage and the chance of circuit interruption. Even with this reduced number of outlets, people using the receptacles should be aware of the kinds of equipment and appliances they connect to the circuit because devices such as fan motors and lighting ballasts may have a high leakage current, leading to nuisance tripping of the interrupter.

Where a GFCI circuit breaker or a feed-through GFCI receptacle is used, long runs of cable after the GFCI should be avoided. A practical rule often used in the industry is to assume that 75 m (250 ft) of two-conductor cable will leak approximately 1 mA. This leakage current will not increase any shock hazard, but it might cause nuisance tripping of the GFCI.

# What are the requirements for bathrooms that contain laundry facilities?

## What are the receptacle requirements?

In some dwellings, laundry facilities are installed in a bathroom for the convenience of the occupant or to make efficient use of space. In these dwellings, such rooms must comply with the receptacle requirements for bathrooms and for laundry rooms and areas. Thus, the minimum number of duplex receptacles is two: one for the washing machine and one located within 1 m (3 ft) of the sink. The receptacle by the sink could also serve as the additional duplex receptacle required in a laundry room or area.

## What are the circuit requirements?

In a combined bathroom and laundry room, all receptacles within 3 m (10 ft) of sinks, bathtubs, or shower stalls must be protected by a ground fault circuit interrupter. However, an exception is made for the receptacle intended for the washing machine or clothes dryer, even when it is within the 3 m (10 ft) range, so long as the receptacle is located

- behind the washing machine or clothes dryer; and
- not more than 600 mm (2 ft) from the floor.

This exception is intended to reduce the likelihood of nuisance tripping of the overcurrent protection for the laundry room's separate branch circuit.

The sink receptacle is normally connected to the separate circuit required to supply the laundry room or area receptacle (see Figure 41). In some cases, the combined bathroom and laundry room is considered to contain two distinct areas, in which case one circuit would be required for the laundry area and another circuit for the bathroom outlets. The local electrical inspector will determine whether or not two circuits are needed.

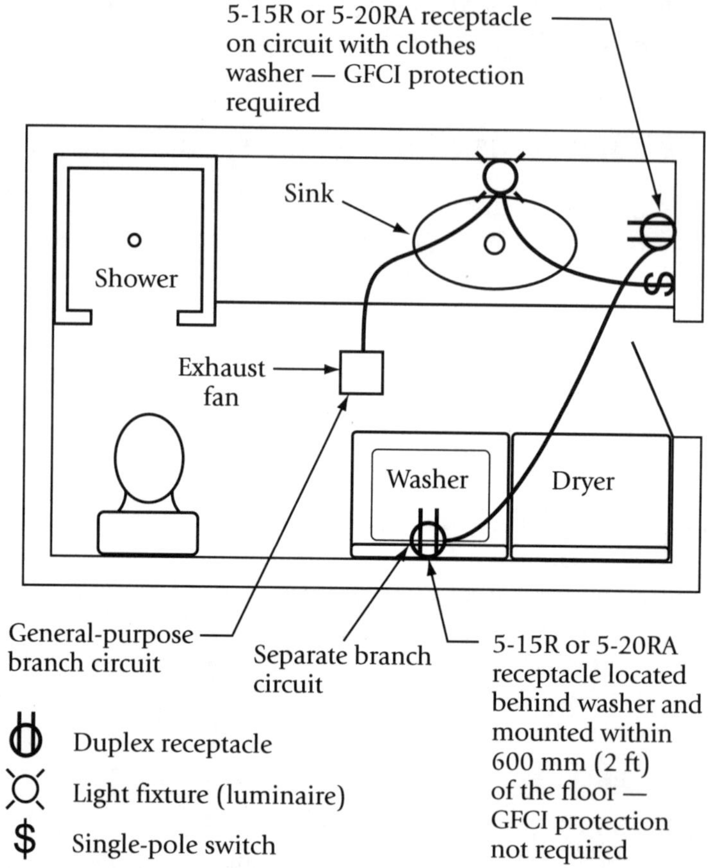


**Figure 41**
**Combined bathroom and laundry room**

# What are the requirements for outdoor receptacles?

## What are the receptacle requirements?

Outdoor receptacles are intended for outdoor appliances and equipment, such as electric lawn and garden appliances, cooking appliances, entertainment appliances, electric snowblowers, car heaters, etc. In single dwellings only, the Code requires that at least one duplex receptacle be installed outdoors, in such a way that it can be reached from the ground or ground level without

- the use of portable ladders, chairs, etc.; and
- having to climb over or remove obstacles.

More specific requirements for the mounting height are not provided, but consideration should be given to possible accumulations of snow that might interfere with easy access to the receptacle.

The Code requires only one outdoor duplex receptacle and sets out no specific requirements for the location of the receptacle. However, it is recommended that an outdoor receptacle be installed at both the front and rear of a dwelling. In colder climates, where block heaters and car warmers are used, a receptacle is often needed near the place where a vehicle is parked. In dwellings with porches, the duplex receptacle required for the porch (see "What are the requirements for receptacles in porches and balconies?" on page 67) could also serve as the required outdoor receptacle. It is also recommended that any outdoor receptacle be controlled by a wall switch located inside the dwelling to prevent unintentional use of the receptacles. See Figure 42.

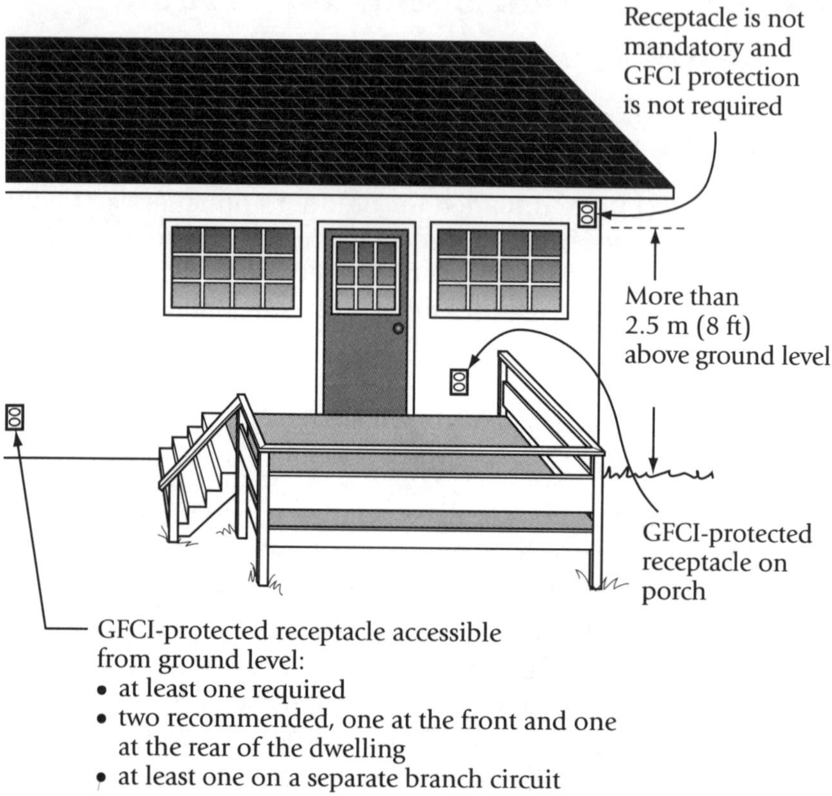


**Figure 42**
**Exterior receptacles**

## What are the circuit requirements?

A single outdoor receptacle must be supplied by a separate branch circuit. Where more than one outdoor receptacle is installed, at least one must be on a separate branch circuit. Additional outdoor receptacles may be connected to that circuit or placed on another general-use branch circuit. All outdoor receptacles within 2.5 m (8 ft) of the ground level must have ground fault protection, provided either by an integral Class A GFCI device or by a Class A GFCI circuit breaker in the panel. See Figure 42.

## What are the requirements for receptacles in unfinished basements?

Residents often wish to put appliances such as freezers or second refrigerators in a basement. Because the basements in many dwellings do not contain finished walls, a specific rule in the Code provides the requirements for these areas.

At least one general-purpose duplex receptacle must be installed in an unfinished basement. The receptacle is a standard outlet and is therefore one of the 12 receptacles and/or outlets that may be connected to a 15 A, 120 V, two-wire general-purpose circuit protected by a 15 A fuse or circuit breaker. In some cases, the receptacle is located adjacent to the panel in the basement, which is convenient for installation. It may also be connected to a branch circuit that was used during the construction of the dwelling or to the same circuit that supplies the basement lights or other outlets in the dwelling.

## What are the requirements for receptacles in garages and carports?

The Code requires one duplex receptacle for each car space in a garage or carport. The garage or carport receptacle(s) must be connected to a separate branch circuit. Lighting fixtures for the area, as well any garage door opening device, may also be connected to this circuit. See Figure 43.

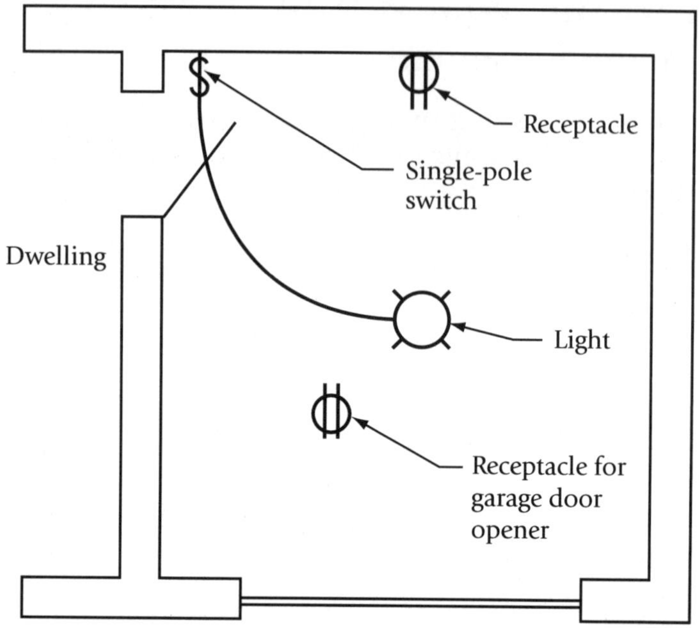


**Note:** *Receptacles and lighting in single-car, attached garages and carports may be connected to the same circuit.*

**Figure 43**
**Receptacles in single-car, attached garages and carports**

In areas with colder climates, where receptacles are needed for more than one vehicle equipped with a block heater or interior car warmer, it may be necessary to provide a separate circuit for each receptacle, depending on the service rating of the devices. Trying to supply more than one heater rated at 300 to 1000 W, for instance, might lead to nuisance tripping of the overcurrent protection, which could easily go unnoticed for some time.

## What are the requirements for receptacles in unattached structures or outbuildings?

In addition to the main dwelling, many properties include unattached structures or outbuildings such as a detached garage, garden shed, pool change room, equipment shed to enclose and protect equipment installed outdoors (e.g., pool, water, and sewage pumps), boathouse, wood or storage shed, children's playhouse, sleeping or housekeeping cabin, pottery or craft studio, etc. These structures usually obtain electrical energy from a panel in the main dwelling.

For most of these structures, the Code has no requirements for receptacles. Receptacles are installed as a convenience for the particular application for which the structure is being used. The following are key considerations in planning the installation of receptacles in such structures:

- Is the receptacle going to create a fire hazard?
- Is the receptacle or the device plugged into the receptacle going to create a shock hazard?
- Will it be possible to maintain the receptacle?
- Is the receptacle protected from damage caused by the environment and physical setting?

Outbuildings or unattached structures are often not built to the requirements of the *National Building Code of Canada* for foundations and structural integrity. This means that the structure may move because of environmental conditions such as weather, frost, ice, and water levels. The structure may also be used only intermittently.

The following are some recommendations for the installation of receptacles in these types of structures:

- The circuits feeding the receptacles or outlets should have clearly labelled and conveniently located switches or circuit breakers that can de-energize the circuits during periods when the receptacle is not in use.
- If the wiring for the circuits from the main dwelling to the outbuilding is overhead, it must
  - have the appropriate ground clearances;
  - use the appropriate type of conductor; and

    - use approved overhead support equipment at both ends.
- If the wiring for the circuits from the main dwelling to the outbuilding is underground,
    - the conductors must be buried at the proper depth;
    - the appropriate type of conductor must be used; and
    - slack must be left in the cable where it enters both structures to allow for movement caused by frost.
- If the receptacle is on the exterior of an outbuilding or is used in structures near water, such as boathouses, pool equipment sheds, etc., the receptacle must be protected by a Class A ground fault circuit interrupter.
- If rodents, such as mice, chipmunks, squirrels, etc., have access to the wiring for the receptacles, a conduit or metal armoured cables should be used to protect the wiring from damage.
- If the outbuilding has sleeping areas or rooms, the receptacles must be protected by arc fault circuit interrupters.

# 5 LIGHTING

## What are the general requirements?

Adequate illumination is an important safety issue in and around a dwelling. Enough lighting is needed both inside and outside to ensure that people have sure footing and are able to identify objects that could cause them to stumble or fall. This is particularly important for cottages where the terrain may be uneven. Outside lighting, while making entry and exit from the residence safer, is also considered a good burglar deterrent and helps people be sure that no one is lurking in the vicinity of the residence.

Requirements for illumination of dwellings are set out in Article 9 of the *National Building Code of Canada*. These requirements are repeated in the *Canadian Electrical Code* to help to ensure that contractors, electricians, and installers comply with all the building and electrical requirements that apply to the installation of lighting in a dwelling.

## What are the lighting requirements for entrances?

The Code requires that lights (or luminaires, a term used to refer generally to any lighting unit with means for direct connection to a branch circuit) be provided at each entrance to a dwelling, with a control inside the dwelling that permits the lighting to be switched on and off. This requirement applies to any entrance, including sliding patio doors, French doors, and exterior doors to basements. The Code does not require illumination at the vehicle entrance to a garage, but it does require a light at the doorway leading from the garage to the dwelling. See Figure 44.

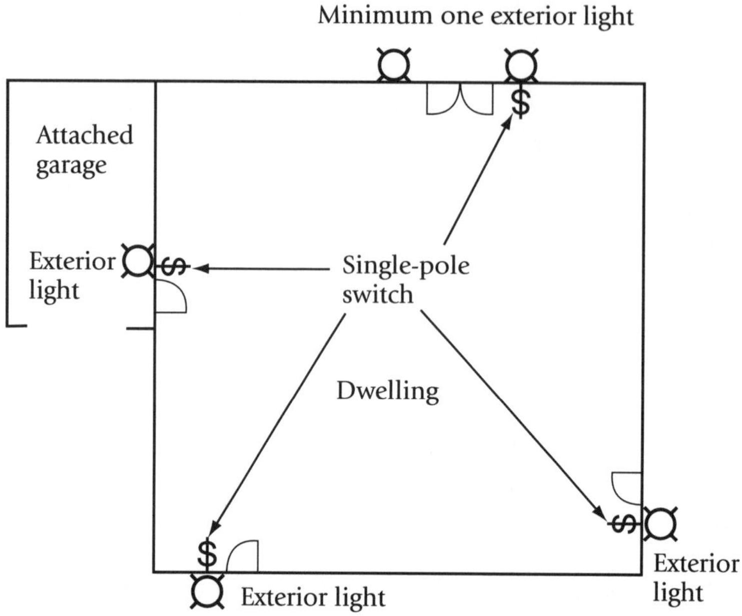


**Figure 44**
**Lighting equipment for entrances**

The entrance lights are controlled by conventional wall switches inside the dwelling. In some installations, lights are also controlled by built-in dusk-to-dawn light sensors and/or motion sensors, in addition to the inside switch. Dusk-to-dawn sensors allow the lights to come on automatically between dusk and dawn. Provided that the inside switch is in the on position, motion sensors turn the lights on when people or vehicles approach the light after dark. (When installing lights with motion sensors, it is a good idea to make sure that the sensitivity of the device can be adjusted so that disturbances caused by weather conditions, such as snow, rain, and wind, or the movement of small animals and birds do not turn the lights on.)

Entrance lighting can also be controlled by electronic programmable switches that turn the lights on between certain hours, varying on and off times at random. In some cases, such programmable switches use a dusk-to-dawn sensor to turn lights on in the evening and shut them off at a prescribed time. These types of controls usually have a manual override feature that allows the inside switch to override the automatic functions.

## What are the lighting requirements inside dwellings?

The Code sets out some basic requirements for lighting in the finished areas inside a dwelling. Kitchens, bedrooms, living rooms, utility rooms, dining rooms, bathrooms, vestibules, and hallways must each have a light controlled by a wall switch. The light may be located on the ceiling of the room or on a wall.

The Code does not specify where switches must be located, but it is an accepted practice for installers, designers, and inspection departments to make sure that switches are not located so that a person has to cross a room or area to turn on the light for that room or area. Similarly, where an area has multiple entrances, 3-way switches are used to allow the operation of the area's lights from any entrance, without requiring a person to move about the area in the dark.

In bedrooms and living rooms that have a receptacle controlled by a wall switch, the requirement for a ceiling or wall light is waived, on the assumption that the room will be sufficiently illuminated by a lamp plugged into the receptacle. Where a duplex receptacle operated by a wall switch (or "switched-duplex receptacle") is used instead of a light outlet and fixture, the receptacle may be included as one of the wall-mounted receptacles required in the usable space of a finished wall (see "What is the minimum number of receptacles required for a finished wall?" on page 61), provided that only half of the receptacle is controlled by the wall switch (see Figures 45 and 46).

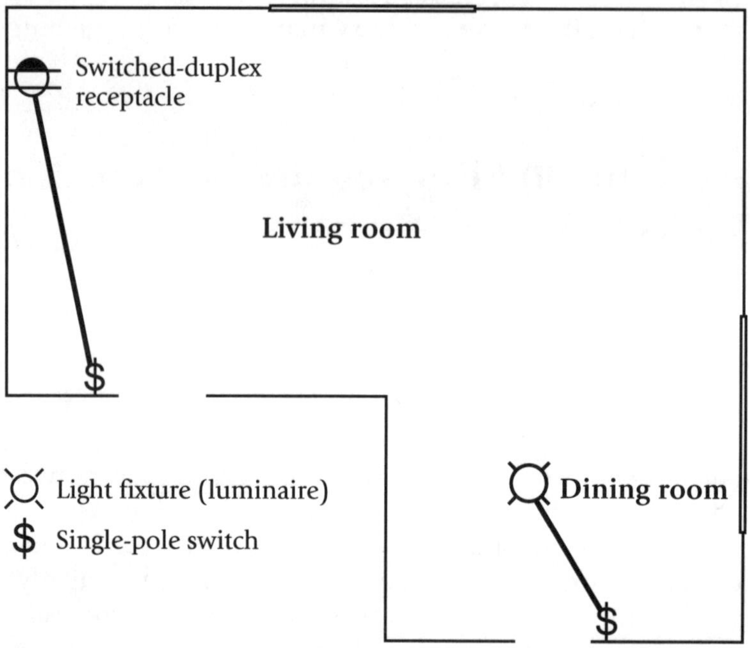


**Figure 45**
**Light switches in combination living room/dining room (1)**

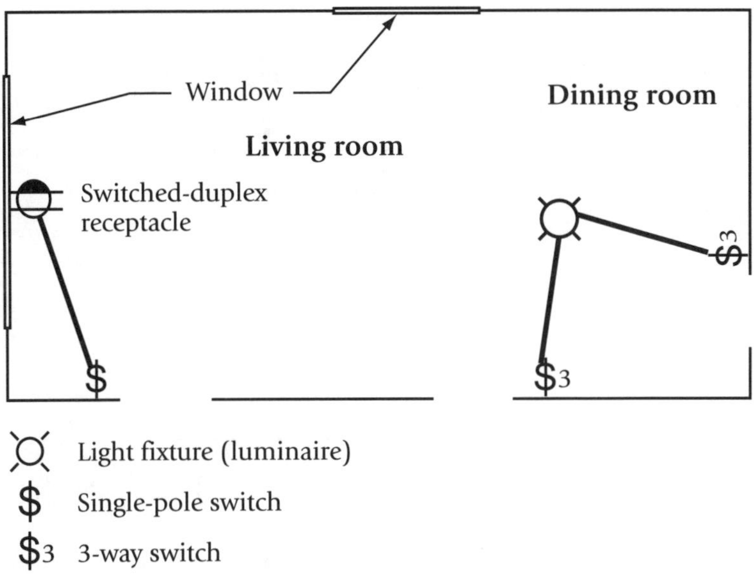


Figure 46
**Light switches in combination living room/dining room (2)**

## What are the requirements for lighting in stairways other than basement stairs?

Stairs must be illuminated to help to ensure the safety of people using them. The Code requires that any stairway having four or more risers be lighted and that the lighting be controlled by 3-way wall switches located at the top and bottom of the stairway (see Figure 47).

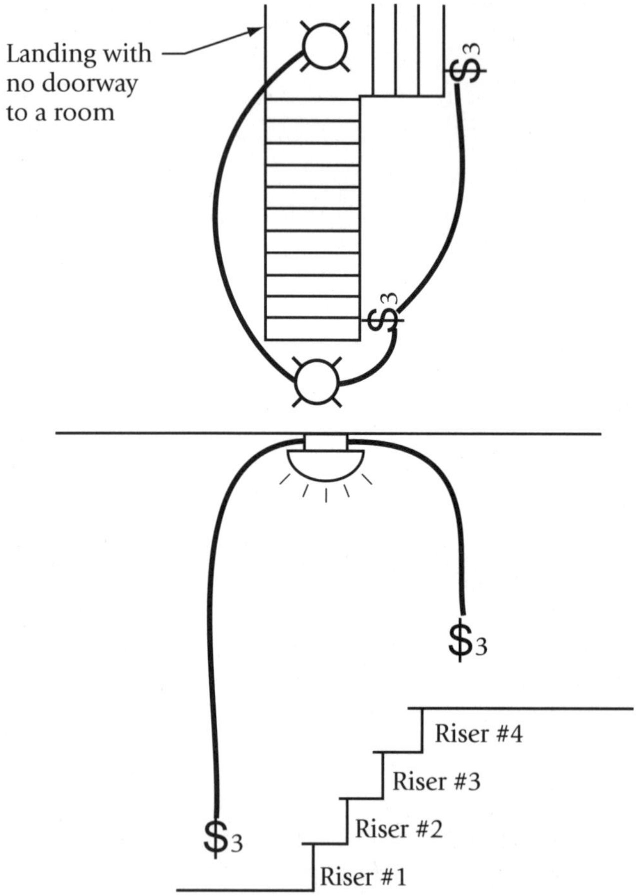


**Note:** *Stairways with four or more risers require 3-way switches at the top and bottom of the stairs.*

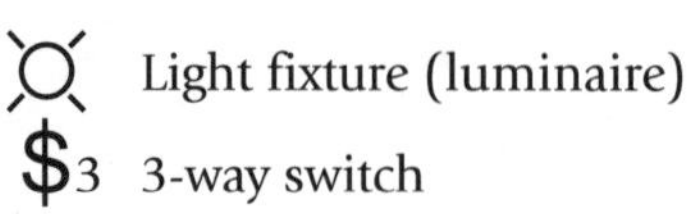


Figure 47

**Switches for stairway lighting**

Stairways with fewer risers are excluded to take into account sunken living rooms or recreation rooms and other minor changes in floor level, where the existing area lighting is adequate.

The number and location of lights needed to provide the required illumination is left to the designer or installer. In general, however, the lighting arrangement should illuminate the stairs at their head and foot and along their entire length, without shadows, to help to ensure the safety of the users.

## What are the lighting requirements for basement stairs?

Basements in dwellings are often left unfinished and do not form a part of a dwelling's general living space. In certain circumstances, the Code permits the lighting for a basement stairway to be controlled by a single switch located at the top of the stairs. The rationale is either that no one will stay in the area very long because it is not finished and offers few amenities, or that anyone using the area cannot leave by another entrance; in both cases, a switch at the bottom of the stairway would be unnecessary. Basement stairway lighting may be controlled by a single switch located at the top of the stairs, provided that the basement does not

- contain a finished area;
- lead to an outside entrance;
- lead to a built-in garage; or
- serve more than one dwelling unit.

See Figure 48.

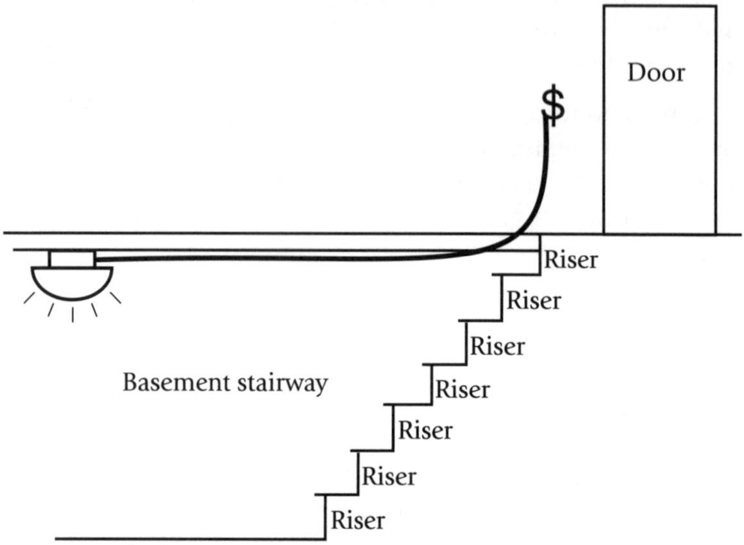


**Note:** *If the basement has a finished area, an entrance to the outside or to a built-in garage, or leads to another dwelling unit, 3-way switches are required at the head and foot of the stairway.*

$ Single-pole switch

**Figure 48**
**Switch for basement stairway lighting**

Note that although a basement stairway may not require a switch at the bottom of the stairs at the time of its original construction, the basement area might be finished at a later date. It is much easier to install a 3-way switch at the original time of construction.

## What are the lighting requirements for unfinished basements?

The Code requires one light for each 30 $m^2$ of floor area (or fraction thereof) in an unfinished basement:

Minimum number of lights = total floor area in $m^2$ / 30

Round any fraction up to the next whole number.

For example, if the floor area of an unfinished basement is 163 $m^2$:

Minimum number of lights = 163 / 30 = 5.46

Therefore, at least six lights are required for the basement.

When the light that provides the required illumination of the basement stairs is located on the ceiling of the unfinished basement and controlled by the wall switch at the top of the stairs, it may be counted as one of the lights required for the unfinished basement area.

While basement lights can be controlled by wall switches, the lights installed in unfinished areas are usually controlled individually by pull-chain switches. This makes it easier to finish areas of the basement later and rearrange the wiring as required.

## What are the requirements for lights in storage rooms?

The Code requires that a light be installed in a storage room, such as a fruit cellar, wine cellar, or pantry. The light may be a pull-chain fixture or controlled by a wall switch.

## What are the requirements for lights in clothes closets?

The Code does not require that lighting fixtures be installed in clothes closets. However, when illumination is installed in a clothes closet, a fire hazard can be created if combustible material on the shelves is piled too close to the closet light fixture. This is especially dangerous if the fixture is the bare bulb type, that is, without a globe or enclosure surrounding the light bulb.

To avoid a fire hazard in a closet, the Code requires that light fixtures installed in a clothes closet be located on the ceiling or on the front wall of the closet above the door. The light fixture must not be located above the shelf, to avoid contact with combustibles. Where a light fixture is located on the trim or sidewall of the closet doorway, it must be a type of light specifically approved for this location. Pendant light fixtures, as well as fixtures without a globe or enclosure surrounding the bulb, must not be used. See Figure 49.

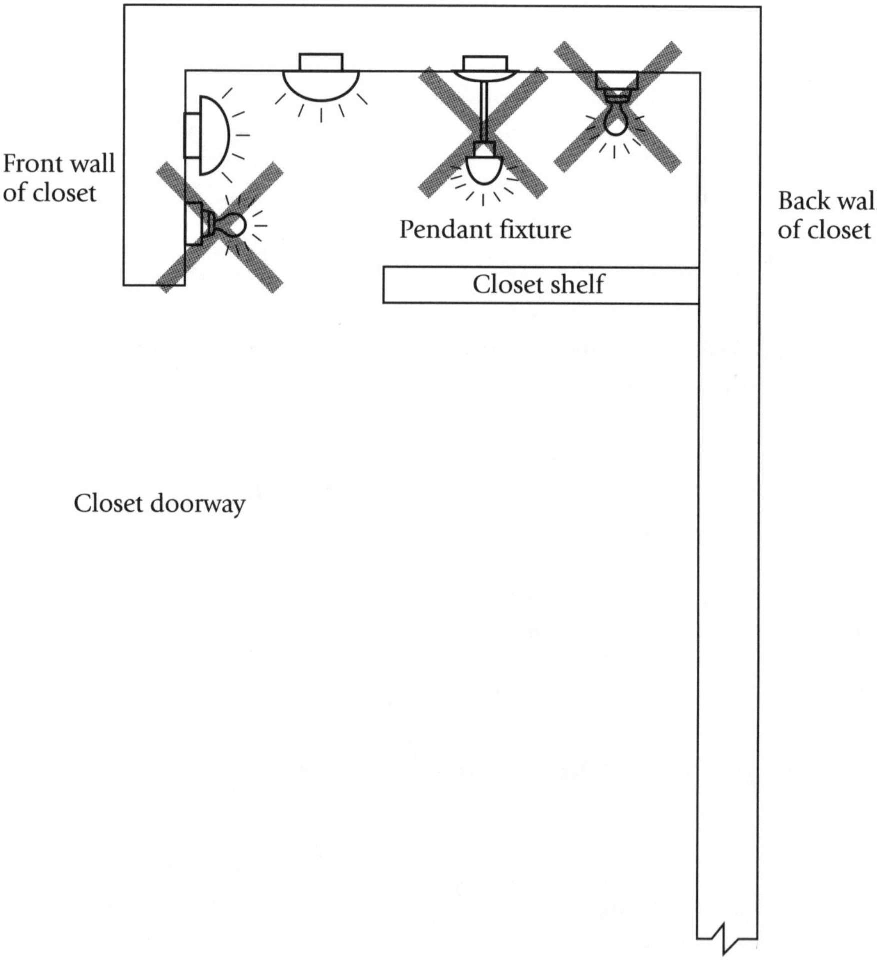


Figure 49
**Light in a clothes closet**

## What are the requirements for lights in garages and carports?

Any attached, built-in, or detached garage or carport must have a light controlled by a wall switch near the doorways (see Figure 50). Lights that are wall-mounted or ceiling-mounted above an area not normally occupied by a parked car may have a built-in switch instead of a wall switch.

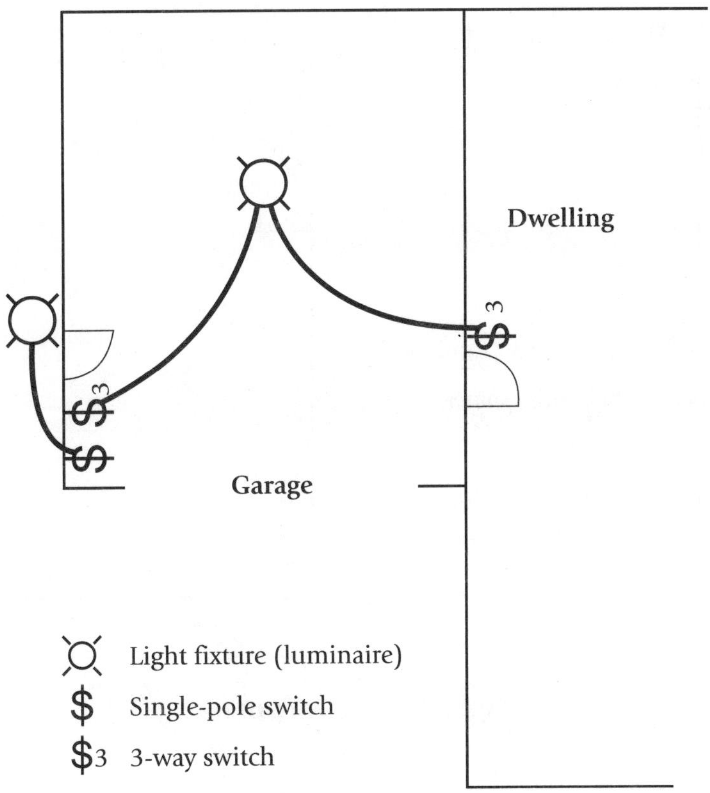


Figure 50
**Lighting for garages**

In dwellings with carports, if the dwelling's entrance lighting (see "What are the lighting requirements for entrances?" on page 107) also illuminates the carport, no additional carport lighting is required (see Figure 51).

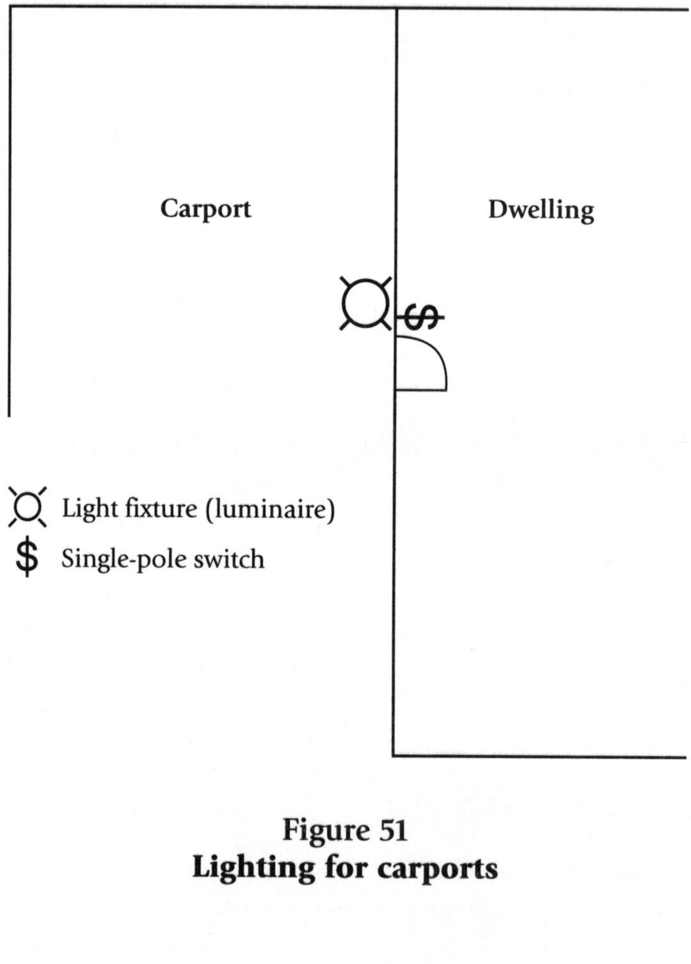


**Figure 51**
**Lighting for carports**

## What are the requirements for lighting in unattached structures or outbuildings?

In addition to the main dwelling, many properties include unattached structures or outbuildings such as a detached garage, garden shed, pool change room, equipment shed to enclose and protect equipment installed outdoors (e.g., pool, water, and sewage pumps), boathouse, wood or storage shed, children's playhouse, sleeping or housekeeping cabin, pottery or craft studio, etc. These structures usually obtain electrical energy from a panel in the main dwelling.

For most of these structures, the Code has no requirements for lighting. Lighting is installed as a convenience for the particular application for which the structure is being used. The following are key considerations in planning the installation of lighting in such structures:

- Is the lighting going to create a fire hazard?
- Is the location of the bulb or fixture going to create a shock hazard?
- Will it be possible to maintain the lighting?
- Is the fixture protected from damage caused by the environment and physical setting?

Outbuildings or unattached structures are often not built to the requirements of the *National Building Code of Canada* for foundations and structural integrity. This means that the structure may move because of environmental conditions such as weather, frost, ice, and water levels. The structure may also be used only intermittently.

The following are some recommendations for the installation of lighting fixtures in these types of structures:

- The circuits feeding the lighting fixtures should have clearly labelled and conveniently located switches or circuit breakers that can de-energize the circuits during periods when the lighting is not in use.
- If the wiring for the circuits from the main dwelling to the outbuilding is overhead, it must
    - have the appropriate ground clearances;
    - use the appropriate type of conductor; and
    - use approved overhead support equipment at both ends.

- If the wiring for the circuits from the main dwelling to the outbuilding is underground,
  - the conductors must be buried at the proper depth;
  - the appropriate type of conductor must be used;
  - slack must be left in the cable where it enters both structures to allow for movement caused by frost.
- If the light fixture is on the exterior of an outbuilding or is used in structures near water, such as boathouses, pool equipment sheds, etc., the fixture must be protected by a Class A ground fault circuit interrupter.
- If rodents, such as mice, chipmunks, squirrels, etc., have access to the wiring for the light fixtures, a conduit or metal armoured cables should be used to protect the wiring from damage.

# 6 Smoke Alarms and Carbon Monoxide Detectors

## What are the requirements for smoke alarms?

### What are the *National Building Code of Canada* requirements?

A smoke alarm is a device that detects the presence of smoke and sounds an alarm at a predetermined level of smoke concentration. It serves as a self-contained fire alarm for a dwelling and alerts only the occupants of that dwelling. For all new construction, the *National Building Code of Canada* requires that smoke alarms conforming to ULC (Underwriters' Laboratories of Canada) Standard CAN/ULC-S531, *Smoke Alarms*, be installed in each dwelling and that these smoke alarms be connected to an electrical branch circuit.

Although it is not required by the *National Building Code of Canada*, it is generally recommended that a dwelling also have a backup alarm that is battery powered, or that one of the smoke alarms connected to an electrical circuit have an integral battery power supply. These alarms will continue to give protection when the power supply to the dwelling is interrupted by a storm, blackout, or other power outages. A battery-powered smoke alarm must be checked periodically, and the battery replaced, to ensure that it continues to provide protection. (A good time to change the batteries in battery-powered smoke alarms is when you change your clocks from standard to daylight saving time, and vice versa.) The requirements for a dual powered smoke alarm (a direct-connected alarm with an integral battery charger) vary. Consult the manufacturer's documentation for instructions on when to change the internal battery.

Smoke alarms should also be tested regularly. Consult the manufacturer of the smoke alarm for the proper procedures for testing. Failure to follow the manufacturer's testing instructions may cause the alarm not to to work properly. For instance, people sometimes use aerosol sprays to test smoke alarms. The spray can leave a film in the smoke alarm, which can make the alarm more sensitive or, by preventing proper air flow, less sensitive.

## What are the location requirements for smoke alarms in a dwelling?

The *National Building Code of Canada* requires that smoke alarms be installed on each floor level, including basements. A floor level is considered to be any floor that is 900 mm (3 ft) or more above or below an adjacent floor level. For example, a two-storey dwelling with a basement requires a minimum of three smoke alarms:

- one for the basement (finished or unfinished) if the basement floor is more than 900 mm (3 ft) below the first floor (some crawl spaces and unexcavated spaces do not require an alarm);
- one for the first floor, given that it is more than 900 mm (3 ft) above the basement; and
- one for the second floor, given that it is more than 900 mm (3 ft) above the first floor.

As well, each bedroom must be protected by a smoke alarm, either inside the bedroom or, if outside, within 5 m (16 ft) of the bedroom door. This measurement is made along the hallways and through the bedroom doorway, not through walls or openings in walls.

Rooms other than bedrooms may be up to 15 m (50 ft) from a smoke alarm. This measurement is made along hallways and going through doorways, not through walls or openings in walls.

Smoke alarms must be installed either on the ceiling or on a vertical wall between 150 mm (6 in) and 300 mm (12 in) from the ceiling. If an alarm is mounted on a vertical wall, the opposite wall must be no more than 1.5 m (5 ft) away. This is to help to ensure that the smoke alarm will detect the smoke from a fire as soon as possible. The wider the area, the more time it takes for smoke to accumulate at the ceiling level — slowing the progression of the smoke down the wall to the smoke alarm — resulting in more time for fire to spread, and possibly preventing a safe evacuation. A wall mounting is not permitted in a hallway or area that is wider than 1.5 m (5 ft). See Figure 52.

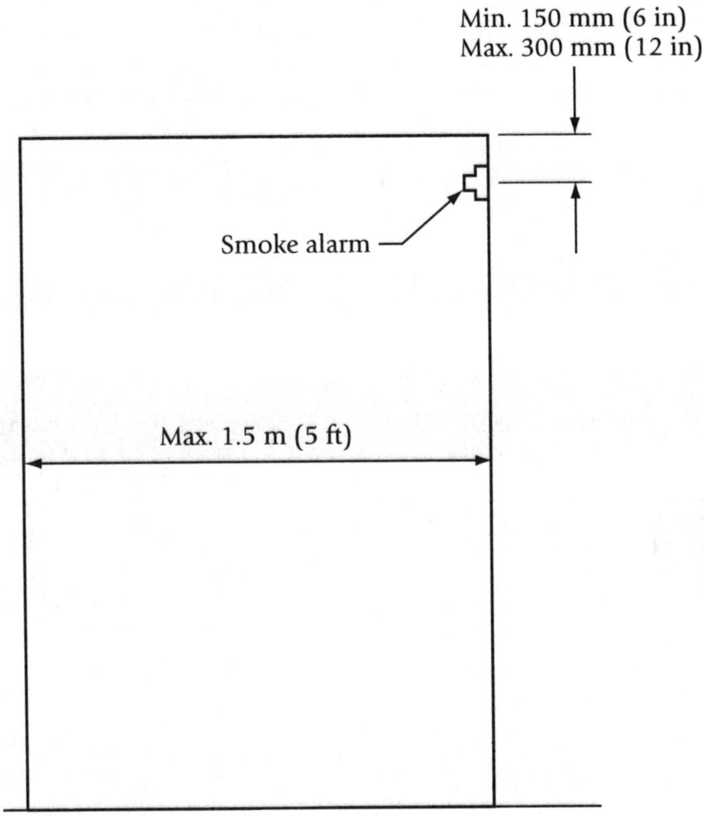


**Figure 52**
**Wall-mounted smoke alarms**

## What are the circuit requirements for smoke alarms in dwellings?

The *Canadian Electrical Code* supplements the *National Building Code of Canada* requirements with specific wiring rules. Smoke alarms may be installed in any circuit that supplies only lighting outlets and in any circuit that has both lighting outlets and receptacles, excluding branch circuits that supply kitchen counter and dining area receptacles, refrigerators, outlets in a garage, and outdoor receptacles. Smoke alarms must not be connected to any circuit that

- is protected by an arc fault circuit interrupter or a ground fault circuit interrupter; or
- has only receptacles connected to it.

The smoke alarm circuit must not be protected by an arc fault circuit interrupter or a ground fault circuit interrupter because circuits with AFCI and GFCI protection are more sensitive than circuits protected by a standard fuse or circuit breaker. As a result, AFCI and GFCI circuits are prone to nuisance tripping, which would compromise the protection offered by smoke alarms.

The advantage of placing smoke alarms on lighting or lighting and receptacle circuits is that if an overcurrent device on the circuit trips, it will be more obvious to the dwelling's occupants because the lights and/or devices on the circuit will not be working, thus ensuring that power will be restored to the circuit and to the smoke alarm more promptly.

To prevent smoke alarms from being inadvertently disconnected from the overcurrent device, the Code requires that no disconnecting means (that is, no switch) be installed between the smoke alarm and the overcurrent device for the branch circuit. This eliminates the possibility of accidentally turning off the alarm.

Where more than one smoke alarm is required because there is more than one floor level or a large floor level, the *National Building Code of Canada* requires that the smoke alarms be interconnected. This means that the actuation of one alarm will cause all the alarms in the dwelling to sound, alerting the occupants no matter where they are in the dwelling. Where multiple alarms are to be installed, it is important to make sure that each device is approved and has been tested for interconnection with the other devices; consult the manufacturer's documentation.

In some circumstances, the interconnection circuits may be Class 2 circuits, in which the use of smaller conductors and cables that are less robust than nonmetallic sheathed cable is permitted. In its general wiring rules for fire alarm systems, the Code requires that either metallic or nonmetallic sheathed cable be used in combustible construction and that metallic sheathed cable be used in noncombustible construction. However, because most dwellings are of combustible construction, the Code generally requires that nonmetallic sheathed cable be used for interconnections. Class 2 wiring methods are thought to compromise the reliability and level of safety provided by interconnected smoke alarms.

### What is the life expectancy of smoke alarms in dwellings?

At present, there is no standard or specific life expectancy for smoke alarms. However, the Canadian Council of Fire Marshals and Fire Commissioners, the National Fire Protection Association in the United States, and many fire departments now recommend that smoke alarms be replaced every ten years.

## What are the requirements for carbon monoxide (CO) detectors?

### What are the *National Building Code* requirements?

For all new construction, the *National Building Code of Canada* requires that carbon monoxide (CO) detectors conforming to CSA Standard CSA 6.19, *Residential Carbon Monoxide Alarming Devices*, be installed in each room containing a solid-fuel-burning appliance that does not incorporate doors that substantially close off the firebox (the chamber in which the fuel burns) when the appliance is in operation. Such appliances include open wood- or coal-burning fireplaces that do not have airtight doors across the hearth, Franklin stoves, and cast-iron pot-bellied stoves. Natural gas and propane gas appliances are not included because they are manufactured and installed in conformance with strict requirements for venting. However, because carbon monoxide gas is produced by many different sources and can reach lethal concentrations unexpectedly, it is a good practice to install an approved CO detector even when the *National Building Code of Canada* does not strictly require one.

A CO detector senses the presence of carbon monoxide, a colourless and odourless gas that can be deadly, and sounds an alarm at a predetermined level of CO concentration. It functions as a self-contained CO alarm system for a dwelling and alerts only the occupants of that dwelling.

As with smoke alarms, the *National Building Code of Canada* requires that CO detectors be permanently connected to an electrical branch circuit. And although it is not a *Canadian Electrical Code* requirement, it is generally recommended that dwellings be provided either with a dual powered CO detector (direct-connected, with a battery backup) or with an extra battery-powered detector. The battery backup or extra detector offers continued protection during power outages. To ensure that the detectors will perform when required, the batteries should be changed regularly (twice a year is recommended).

CO detectors should also be tested regularly, in accordance with the manufacturer's instructions for testing. It should be kept in mind that the test button on the detector only tests the electrical connection, not the ability of the device to sense the level of carbon monoxide in the air. There is no product readily available to test the sensitivity of a CO detector. The use of an unapproved material to test a CO detector can damage the device because the reaction between the material and the detector's sensing elements can make the detector either too sensitive or not sensitive enough.

## What are the location requirements for CO detectors in a dwelling?

The *National Building Code of Canada* requires that CO detectors be permanently connected to an electric circuit and installed either on or near the ceiling in each room that contains a solid-fuel-burning appliance. The *National Building Code* does not set out any requirements for wall mountings, but the requirements for wall-mounted smoke alarms are usually followed, with the detectors installed between 150 mm (6 in) and 300 mm (12 in) from the ceiling when the width of the area or hallway is 1.5 m (5 ft) or less. In wider hallways or areas, ceiling mountings are usually preferred.

## What are the circuit requirements for CO detectors in dwellings?

The *Canadian Electrical Code* makes no specific wiring requirements for CO detectors. However, installers and contractors usually follow the Code requirements for smoke alarms.

Consequently, CO detectors are installed in a circuit that supplies only lighting outlets or that has both lighting outlets and receptacles. CO detectors must not be connected to branch circuits that supply kitchen counter and dining area receptacles, refrigerators, outlets in a garage, or outdoor receptacles, or to any general-use branch circuit that

- is protected by an arc fault circuit interrupter or a ground fault circuit interrupter; or
- has only receptacles connected to it.

As with smoke alarms, the exclusion of branch circuits protected by arc fault circuit interrupters or ground fault circuit interrupters helps to ensure that interruptions of power to the CO detector's circuit are noticed and repaired promptly. It is also standard practice to make sure that there is no disconnecting means (no switch) between the CO detector and the overcurrent device for the branch circuit so that the detector cannot be turned off by accident.

The *National Building Code of Canada* requires in general that a CO detector be equipped with an integral alarm that meets the audibility requirements of CSA Standard CSA 6.19, *Residential Carbon Monoxide Alarming Devices*. As an alternative, a CO detector without an integral alarm, or having an alarm that does not meet the audibility requirements, must be interconnected with the dwelling's smoke alarms, so that the actuation of any CO detector or smoke alarm causes all the detectors and alarms in the dwelling to sound. It is good practice to interconnect the CO detectors with the smoke alarms, whether or not it is mandated by the *National Building Code of Canada*. It is essential to make sure that each CO detector and smoke alarm that is to be interconnected is approved and has been tested for interconnection.

The interconnection should comply with the general wiring rules for fire alarm systems. The interconnection circuit may be a Class 2 circuit, but the Code requires the wiring methods in combustible construction to use either metallic or nonmetallic sheathed cable; in noncombustible construction, metallic sheathed cable must be used. The installation of nonmetallic sheathed cable and wiring methods must conform to the Code's rules for nonmetallic sheathed cable. This ensures that the wiring method does not compromise the required reliability and level of safety of the interconnected devices.

## What is the life expectancy of CO detectors in dwellings?

Unlike smoke alarms, CO detectors have a relatively short life span. CO detectors certified under CSA Standard CSA 6.19 have a life expectancy of at least three years and must be replaced once they have passed their rated life expectancy. A CO detector approved under CSA Standard CSA 6.19 either has its expiry date marked on the nameplate or is equipped with an automatic "kill" switch that de-energizes the detector and sounds an alarm at the end of its life cycle.

# 7 Major Appliances and Electrical Equipment

## What special wiring needs commonly exist in dwellings?

Most dwellings contain a number of heavy-duty appliances and pieces of electrical equipment that make considerable electrical demands. These include direct-connected kitchen equipment, water heaters, space heaters and furnaces, air conditioners and heat pumps, pool and spa equipment, central vacuum systems, and large appliance motors. In many cases, there are special receptacle, circuit, and overcurrent protection requirements for this equipment and the areas in which it is installed. This chapter examines the requirements that are most likely to apply in residential installations.

## What are the requirements for direct-connected kitchen equipment?

### What are the requirements for a built-in oven?

When a built-in oven, either gas or electric, is installed in a kitchen, it is connected directly to a separate branch circuit; there is no receptacle. Circuit requirements vary depending on the model of oven that is being installed. Conductors and overcurrent protection are determined by the oven's wattage, voltage, and current ratings (see "How are branch circuits sized?" on page 50).

## What are the requirements for a built-in countertop range?

Built-in countertop ranges, whether gas or electric, are connected directly to a separate branch circuit. The range's wattage, voltage, and current ratings determine the sizing of the circuit's conductors and overcurrent device (see "How are branch circuits sized?" on page 50). Always check the requirements of the specific model for accurate installation and electrical requirements.

## What are the requirements for a built-in oven and built-in countertop range treated as a single unit?

The Code permits two or more built-in cooking units to be connected directly to one separate branch circuit. The conductor size and overcurrent protection for such a circuit are based on the sum of all the ratings of the built-in cooking units (see "How are branch circuits sized?" on page 50). See Figure 53.

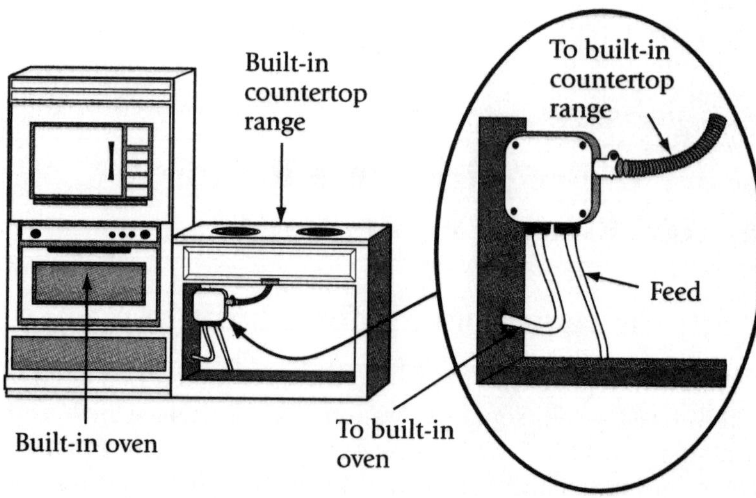


Figure 53
**Branch circuit connection for built-in ovens and countertop ranges**

# What are the requirements for electric water heaters?

## What are the different kinds of electric water heaters?

There are two kinds of electric water heaters:

- storage-tank water heaters; and
- in-line electric water heaters.

A storage-tank water heater, the type most commonly used in the residential market, consists of a storage tank with a capacity of 40 to 60 gallons and two electric elements located inside the storage tank, near the top and the bottom. Each element is controlled by a thermostat that energizes the element when the temperature of the water adjacent to it drops below the desired temperature. The water heater is designed so that only one element can be energized at a time, through the use of a flip-flop switch adjacent to the upper thermostat. See Figure 54.

An in-line water heater does not have any storage capacity. It has one element that heats water on demand as it flows through the heater.

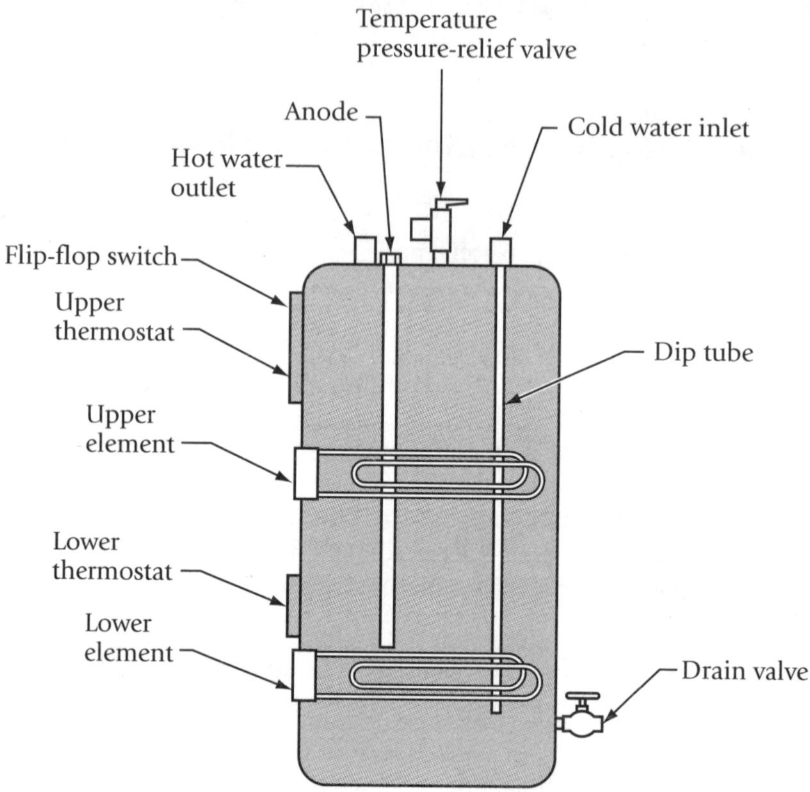


Figure 54
**Storage-tank water heater**

## What are the requirements for an electric storage-tank water heater?

The two elements in a storage-tank water heater often have the same wattage rating. In some designs, the top element has a larger rating than the bottom element. Typical configurations are shown in Table 6.

| Capacity, gal | Top element, W | Bottom element, W |
|---|---|---|
| 40 | 3000 | 3000 |
| 40 | 3000 | 1000 |
| 60 | 4500 | 4500 |
| 60 | 4500 | 1500 |

**Table 6**
**Element ratings in storage-tank water heaters**

Unlike other types of electric heating appliances, storage-tank water heaters must be connected directly to a separate branch circuit. The size of conductor used to supply the water heater and the rating of the overcurrent device needed to protect the water heater can be determined from Table 7 or 8, using the voltage rating (120 V or 240 V) and wattage rating shown on the heater's nameplate. (The rating of the water heater in watts is usually the rating of the largest element because both elements are not used at the same time.)

| Maximum connected heating load, W | Minimum conductor size using NMD90 copper cable | Maximum fuse or circuit breaker rating, A |
|---|---|---|
| up to 1440 | #14 | 15 |
| up to 1920 | #12 | 20 |
| up to 2880 | #10 | 30 |
| up to 3840 | #8 | 40 |
| up to 5760 | #6 | 60 |
| up to 9600 | #3 | 100 |
| up to 12 000 | #2 | 125 |

**Note:** *Where more than one direct-connected electric heating appliance is connected to a branch circuit, the rating of the fuse or circuit breaker must not exceed* 30 A.

**Table 7**
**Sizing chart for 120 V direct-connected electric heating appliances**

| Maximum connected heating load, W | Minimum conductor size using NMD90 copper cable | Maximum fuse or circuit breaker rating, A |
|---|---|---|
| up to 2880 | #14 | 15 |
| up to 3840 | #12 | 20 |
| up to 5760 | #10 | 30 |
| up to 7680 | #8 | 40 |
| up to 11 520 | #6 | 60 |
| up to 19 200 | #3 | 100 |
| up to 24 000 | #2 | 125 |

**Note:** *Where more than one direct-connected electric heating appliance is connected to a branch circuit, the rating of the fuse or circuit breaker must not exceed* 30 A.

**Table 8**
**Sizing chart for 240 V direct-connected electric heating appliances**

More than one direct-connected electric heating device may be connected to a branch circuit. For example, the same circuit might supply two or more electric baseboard heaters, two electric water heaters, or a fan-type electric heater and electric baseboard heaters. In such a case, the Code limits the total equipment load on the circuit to 5760 W, which is 80% of the total load that theoretically could be placed on the circuit:

W = V × A

= 240 V × 30 A

= 7200 W

and

7200 W × 0.80 = 5760 W

As well, the maximum rating of the overcurrent protection is restricted to 30 A. These limits on equipment load are based on factors such as continuous or non-continuous load, continuous operation rating of the overcurrent device, and the wiring method used.

In most residential installations, electric storage-tank water heaters are wired with nonmetallic sheathed cable. The Code requires that nonmetallic sheathed cable be protected from damage when it is run within 1.5 m (5 ft) of the floor, the height at which the connection is usually made to the water heater. Where the size and/or location of the water heater exposes the wiring to damage, nonmetallic sheathed cable may still be used, provided it is enclosed in 16 mm (½ in) metal flexible conduit supported by a wire or chain or by a short piece of wood run from the ceiling. As an alternative, armoured cable may be used because its spiral-wrapped metal jacket provides extra mechanical protection.

## What are the requirements for an electric in-line water heater?

The conductor, circuit, and overcurrent protection requirements are the same for in-line water heaters as for storage-tank water heaters. Tables 7 and 8 show the minimum conductor size and maximum fuse or circuit breaker rating for a range of equipment loads, as determined by the heater's nameplate voltage and wattage ratings. An in-line water heater must be connected directly to a separate branch circuit.

In-line water heaters heat the water on demand before it flows through the hot water faucet. They are usually installed near the sink, inside a cupboard beneath the sink, in a ceiling space above the sink, or in a similar location. They may be wired with nonmetallic sheathed cable, which must be protected wherever it runs within 1.5 m (5 ft) of the floor or through any area where it may be exposed to physical damage. Installation in a cupboard or a cabinet cannot be relied on to provide protection to nonmetallic sheathed cable because it may be exposed to physical damage whenever the cupboard or cabinet is opened and objects are put in or taken out. As with storage-tank water heaters, armoured cable can be used where the location of the water heater would expose nonmetallic sheathed cable to damage. Otherwise, proper protection must be provided, such as enclosure in 16 mm (½ in) metal flexible or liquid-tight flexible conduit or attachment to a support chain.

## What are the requirements for space-heating equipment?

### What are the preliminary considerations?

The first decision regarding space-heating equipment is choosing the fuel it will use: electricity, natural gas, propane, oil, or in some cases, coal, wood, or pellet fuels. This choice is usually made by considering the following factors:

- the availability of the fuel;
- the cost of the fuel and equipment;
- the efficiency of the fuel source;
- maintenance costs;
- the space required for equipment and ductwork; and
- associated functions and equipment, such as
  - electronic air cleaners;
  - humidifiers;
  - air conditioners (heat pumps);

- individual room thermostats (rather than one temperature control for the whole dwelling); and
- heat exchangers.

The size of the space-heating equipment selected for a whole dwelling or a single room or area is best determined by making a heat loss calculation, which takes into account

- insulation types and ratings (R values) for the dwelling;
- types and sizes of windows and exterior doors;
- shapes and ceiling heights of rooms or areas;
- size and area of exterior walls; and
- average high and low outdoor temperatures in the area (as provided by the *National Building Code of Canada* or the Meteorological Service of Canada).

Heating equipment manufacturers and installers, dwelling designers, and mechanical or electrical contractors are all familiar with heat loss calculations.

## What are the requirements for natural gas, propane, and oil furnaces?

Residential furnaces typically have outputs of up to 400 000 BTU per hour (BTUh). Special requirements may apply to space-heating equipment with higher outputs; both the Code and the local electrical authority should be consulted when such equipment is installed.

Furnaces rated up to 400 000 BTUh must be connected directly to a single branch circuit that is used for no other purpose. For most forced-air residential furnaces, a 15 A, 120 V circuit is adequate. When determining the circuit requirements for other types of furnaces, such as hot water systems (boiler type), heat pumps, and unit heaters, always consult the manufacturer's specifications.

Each furnace must be supplied with a disconnecting means that cuts off the electrical supply to the furnace to allow maintenance and servicing. Usually this is a switch (sometimes called an isolating switch). In most cases, a general-purpose switch can be used. It must be located between the furnace and the point of entry into the area in which the furnace is installed. It must not be located on the furnace or in a place that can be reached only by passing close to the furnace. The Code also requires that the switch be marked to indicate that it controls the furnace, to ensure that occupants are aware of its function. As an alternative, the Code also permits the branch circuit breaker in the distribution panel to be used as the disconnecting means, as long as the panel is located between the furnace and the point of entry to the furnace area.

Most residential furnaces have a motor that is rated at less than 1 HP (horsepower), which is used to circulate air or water. It can use a general-purpose switch as the required disconnecting means. However, where a motor has a rating greater than 1 HP, the disconnecting means or switch must have a horsepower rating equal to or greater than that of the motor to prevent damage to switch contacts and the possibility of the premature failure of the switch.

When a motor starts from zero rotation (locked rotor condition), it can draw as much as six times more current than the motor's full load current — the current a motor draws when it is turning at its name-plate speed (expressed in revolutions per minute, or rpm) and delivering its nameplate power to the load it is rotating (expressed in horsepower, or HP).

The full load current of a 2 HP motor at 240 V, single-phase, for example, is approximately 12 A. But when this motor is started at its rated line voltage, it will draw approximately 72 A from the line conductors feeding it. This current draw, which gives the motor the extra torque (starting torque) it needs to start rotating from a standstill, is called a "starting surge" and lasts for a very brief period of time. The contacts of switches or disconnecting means used to start or stop such a motor are designed to take this surge current into account. In a general-purpose switch or a switch not rated for the application, the contacts in the switch could deteriorate before the rated life expectancy of the switch, either failing to make contact (preventing the motor from operating) or fusing together (preventing the switch from disconnecting the power supply to the motor).

Starting surges may cause the lights in a room or area to dim. The high starting current required by the motor takes current away from lighting circuits, dimming the lights briefly before achieving full power when the motor reaches its operating speed. Apart from the potentially annoying dimming, these starting surges usually do not lead to any problems with lighting. However, they can cause problems with sensitive electronic equipment (such as computers) that requires a very steady and stable supply of current — another good reason to keep a furnace motor on a separate branch circuit.

While equipment associated with the furnace, such as electronic air cleaners and humidifiers, is not required to be connected to the separate branch circuit for the furnace, it often is because of the typically small loads it carries. However, any equipment that is essential for the safe operation of the furnace, such as a make-up air fan, a fuel pump, or a circulating pump or fan, must be connected to the furnace circuit.

In most residential installations, furnaces are wired with nonmetallic sheathed cable from the distribution panel to the disconnect switch. The connection for the furnace is usually close to the floor, between 600 mm (2 ft) and 900 mm (3 ft) above the floor, where nonmetallic sheathed cable must be protected from damage. One way to provide this protection is to run armoured cable from the disconnect switch to the furnace connection. Another method is to run nonmetallic sheathed cable from the disconnect switch to the furnace connection and to enclose the cable between the furnace connection and a point at least 1.5 m (5 ft) from the floor in flexible metal conduit. The conduit must be at least 16 mm (½ in) in diameter and be connected to the furnace connection box by a box connector approved for use with flexible metal conduit.

Cable running to the isolating switch and to the furnace must not be supported by gas piping or water piping. The furnace and its ductwork — except for any covers or panels used for access — may be used to support cable (see Figures 55 and 56).

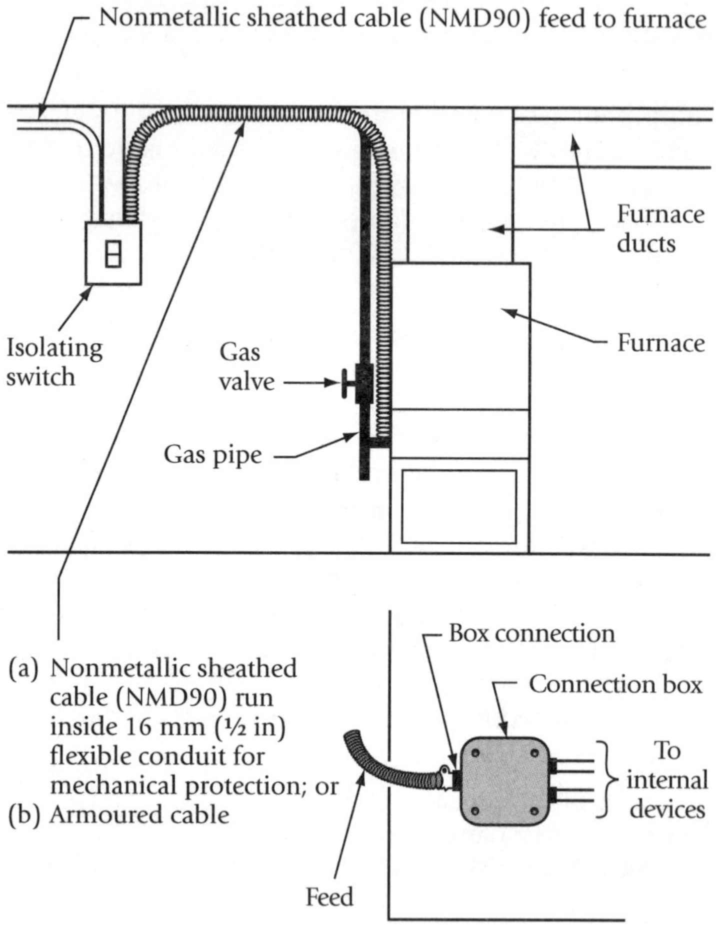


**Figure 55**
**Furnace connection (1)**

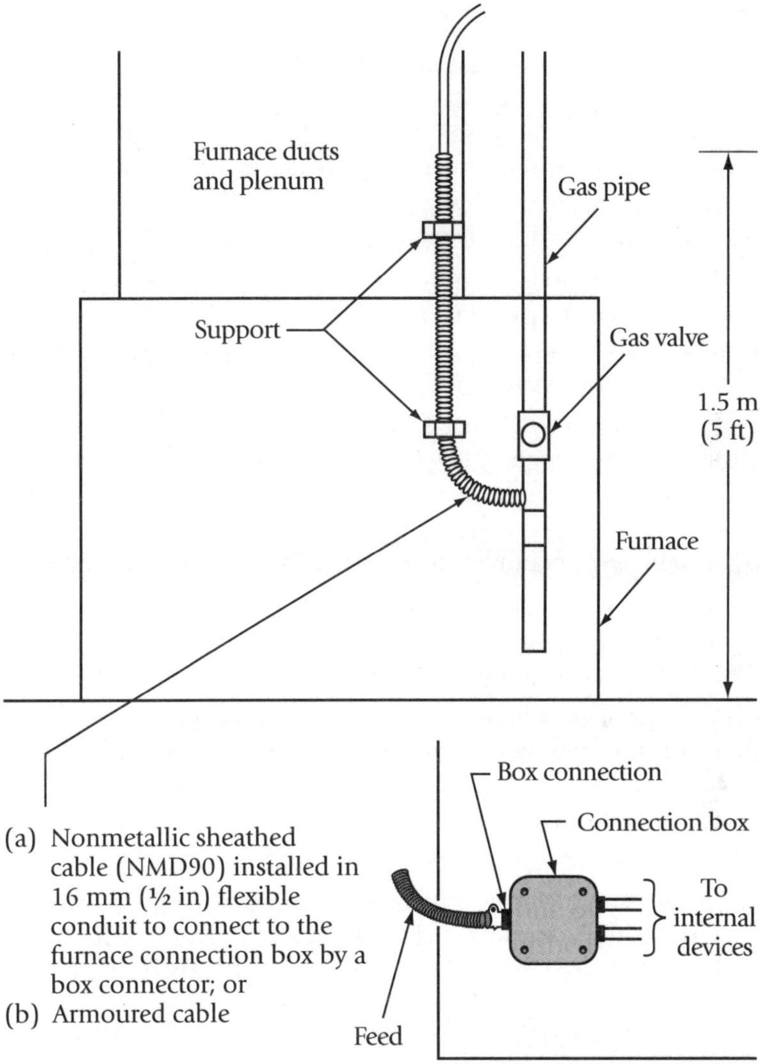


**Figure 56**
**Furnace connection (2)**

## What are the requirements for electric furnaces?

The first consideration in installing an electric furnace is to determine what size of furnace will be needed to heat the dwelling. Here again, a heat loss calculation is required, whether the plan is to use a single electric furnace or individual electric room heaters. The electrical supply authority also needs this calculation because it helps them forecast the loads that will be put on their system.

Electric furnaces consist of banks of 5 kW electric heaters. A 10 kW furnace has two banks of 5 kW heaters, a 15 kW furnace three banks, a 20 kW furnace four banks, and so on, up to a maximum of 30 kW (six banks) for residential furnaces. When the thermostat requires the furnace to be switched on, not all the banks of 5 kW heaters are turned on. The heater banks are controlled by an internal sequence control circuit that activates the number of heater banks required to bring the plenum chamber up to the desired temperature within a specific time and to maintain the temperature as long as the thermostat is calling for heat.

Electric furnaces must be connected directly to a three-wire circuit. The electric heating elements require line-to-line voltage (usually 240 V), while the fan motor and controls usually require line-to-neutral voltage (usually 120 V). The minimum size of the conductors, which must be installed in a raceway or provided as a cable, and of the overcurrent protection required for the circuit, is based on the furnace's kW rating (see Table 8 on page 134), which appears on the furnace nameplate and in the manufacturer's documentation.

Electric furnaces in dwellings must also be supplied with a disconnecting means. Usually, the circuit breaker in the panel serves to cut off electrical supply to the furnace. When fuses, not circuit breakers, are used as the overcurrent protection, a switch that will open all live conductors to the furnace must also be provided, ahead of the fuses, because it is not a safe practice to remove fuses that are under load or energized. Such a switch, usually called a fusible disconnect switch, comes in standard ratings of 30, 60, 100, 200, 400, and 600 A.

The thermostat for an electric furnace usually switches a relay that, in turn, controls the operation of the furnace. While there are line voltage thermostats that switch the total line current for a heating appliance directly, in residential installations these thermostats are restricted by the Code to a 30 A rating. Electric furnaces usually draw more than 30 A, so line voltage thermostats are seldom used in dwellings. Instead, large loads are controlled indirectly by a thermostat that energizes or de-energizes a coil in a switch, according to whether or not heat is required. The switch then energizes or de-energizes the electric furnace. Relay thermostats are smaller and cheaper than line voltage thermostats, and because they are usually classified as extra-low-voltage equipment (30 V or less), less expensive and less demanding Class 1 or 2 wiring methods may be used between the thermostat and the contactor. Thermostats have varying wiring requirements, and the manufacturer's installation instructions should always be consulted.

If you are planning on using individual electric heaters for each room, a thermostat will be required in each room to control the heater. The thermostat can be built into the heater or wall mounted separately. Because the electric heating circuits are 20 A or less, a line voltage thermostat can be used.

Electronic thermostats can be either the line voltage type (up to 20 A at 240 V) or the extra-low voltage type that switches a relay. Electronic thermostats have many features that can help to save energy costs, such as setting the temperature back in parts of the dwelling that are not being used and adjusting the temperature to accommodate different needs for weekdays and weekends.

# What are the requirements for air conditioners?

## What are the preliminary considerations?

Two kinds of air conditioners are commonly used in dwellings: individual air-conditioning units that are placed in the room where the cooling is required, either in the window or in a specially designed sleeve in an exterior wall, and central units, which use a forced-air furnace and its associated ducts to circulate cooled air throughout the dwelling.

Both individual units and central air-conditioning systems use a compressor to compress Freon® gas in a condenser coil, a process that gives off heat. (That is why the compressor is installed outdoors.) The compressed Freon® is then pumped into an evaporator coil in the equipment indoors, where it expands and absorbs heat from air circulated through the evaporator coil, cooling the air. The Freon® is then pumped back to the compressor, where it is again compressed in the condenser coil, giving off the heat it absorbed from the inside air.

In equipment involving a fan and a motor, or a pump and a motor, the motor drives the equipment through a belt and pulley or a flexible direct connection. In an air-conditioning compressor, however, the motor and the compressor parts are combined on the same shaft in a hermetically sealed enclosure that contains the Freon® gas.

Air-conditioning equipment for residential installation is selected on the basis of the following considerations:

- the type of energy required: electricity for standard air-conditioning compressors or natural gas for absorption-type air-conditioning compressors;
- the cost of the energy and equipment;
- the efficiency of the equipment;
- maintenance costs;
- the space required for the equipment and ductwork; and
- associated functions and equipment, such as
    - electronic air cleaners;
    - dehumidifiers;
    - heat pumps;
    - individual room units (rather than a central unit for the whole dwelling); and
    - heat exchangers.

Air-conditioning equipment, whether intended for a whole dwelling or a single room or area, is sized by making a heat gain calculation. Like a heat loss calculation, this takes into account

- insulation type and ratings (R values) for the dwelling;
- types and sizes of windows and exterior doors;
- shapes and ceiling heights of rooms or areas;
- size and area of exterior walls; and
- average high and low outdoor temperatures.

## What are the requirements for individual air-conditioning units?

Individual units are usually connected to the electrical system by a power supply cord. This allows the unit to be connected or disconnected from the electrical system for servicing or storage in the off season.

It is generally recommended, although not required by the Code, that an individual unit be placed on a separate branch circuit used only for the air-conditioning unit. These units cycle on and off to meet the user's temperature requirements and require a large surge of current during the starting process. The starting surge for an air-conditioning compressor is the same as for a motor, although it may last longer because the combined compressor and motor typically impose a larger load. If there are other loads on the same circuit, nuisance tripping of the fuse or circuit breaker may occur.

The sizing of the conductors and overcurrent protection for the air conditioner circuit is based on the load characteristics of the air conditioner, which vary from model to model. Manufacturer's instructions should specify the circuit requirements, and nameplate information can also be used to size the circuit in accordance with the rules for any plug-in appliance (see "How are branch circuits sized?" on page 50). Individual units rated up to 8000 BTU can be operated on a 15 A circuit. If fuses rather than circuit breakers are used to provide the branch circuit overcurrent protection, Type D fuses should always be used to avoid nuisance tripping. Type D fuses have time-delay characteristics that can accommodate the longer duration of the starting surge associated with air-conditioning compressors.

## What are the requirements for central air-conditioning units?

Central air-conditioning units use the dwelling's forced-air furnace fan to circulate cooled air through the dwelling's ductwork. An outside compressor circulates a refrigerant through a coil placed in the furnace plenum to cool the air moving through the furnace. The thermostat and the control circuit prevent the heating and cooling systems from being on at the same time.

The size of central unit required for a dwelling is determined by making a heat gain calculation in BTU for the dwelling and selecting a unit with the required characteristics, as stated on the compressor's nameplate or in the manufacturer's documentation. Central units for dwellings usually require a 240 V, single-phase circuit, with current ratings varying from 8 to 30 A.

The compressor rating determines, in turn, the circuit requirements for the unit. The minimum conductor size and the maximum fuse or circuit breaker can usually be found on the nameplate of the compressor unit; alternatively, the nameplate information can be used, in conjunction with Table 9, to establish the required conductor size and overcurrent protection. When fuses are used as the overcurrent protection in the panel, they must be installed in a fusible disconnect switch that has a horsepower rating that is at least equal to the compressor unit's horsepower rating, which can also be found on the compressor's nameplate or in the manufacturer's documentation.

The compressor unit is located outdoors in a well-ventilated area. It is usually quite heavy and, rather than being placed directly on the ground, it should be placed on a pad [usually 50 mm (2 in) patio slabs], mounted on a bracket attached to an exterior wall, or otherwise kept out of direct contact with the earth.

The power supply line for the outdoor compressor is run directly from the panel inside the dwelling. The usual method is to run the power lines for the outdoor compressor out through the hole made in the exterior wall for the refrigeration lines. While the power lines usually follow the same route as the refrigeration lines, they must not be supported by the refrigeration lines. Instead, the Code requires that they be independently supported, so that one set of lines can be serviced without disturbing the other. Most contractors and installers use a cable that is approved for use indoors, such as NMD90 (a nonmetallic sheathed cable), and run it through a weatherproof-rated raceway, such as liquid-tight flexible conduit, for the portion of the line that is outside the dwelling.

The power line to the compressor must first go to a weatherproof-rated disconnect switch, which must be mounted within 3 m (10 ft) of the compressor and in the direct line of sight of any person working on the compressor (see Figure 57). This switch must have a horsepower rating at least equal to the horsepower rating of the compressor unit. The disconnect switch is sometimes called an isolating switch because it opens all the live conductors to the equipment, allowing a person to service the equipment without having to worry about live or energized conductors or unexpected start-ups, both of which could cause serious injury or even death.

When both heating and air conditioning are installed in a dwelling, a combination thermostat is used to control both systems. The manufacturer's wiring instructions should be consulted to determine the size and number of conductors that run from the thermostat to the furnace and from the furnace to the outdoor compressor.

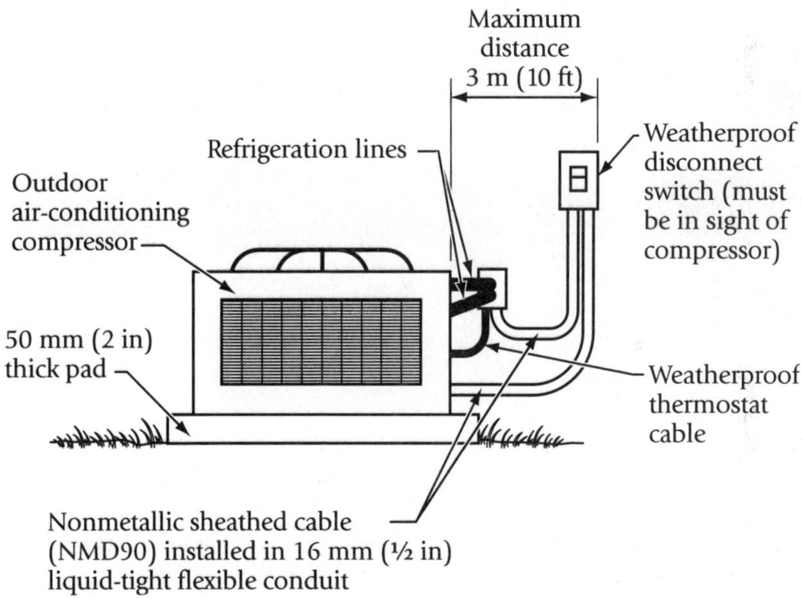


**Figure 57**
**Air-conditioning compressor connections and disconnect switch**

| Compressor rating, A | Minimum conductor size, AWG | Maximum overcurrent device size, A |
|---|---|---|
| 1–6 | #14 | 15 |
| 7 | #14 | 20 |
| 8 | #14 | 20 |
| 9 | #14 | 25 |
| 10 | #14 | 30 |
| 11 | #14 | 30 |
| 12 | #14 | 35 |
| 13 | #12 | 35 |
| 14 | #12 | 40 |
| 15 | #12 | 45 |
| 16 | #12 | 45 |
| 17 | #10 | 50 |
| 18 | #10 | 50 |
| 19 | #10 | 50 |
| 20 | #10 | 60 |
| 21 | #10 | 60 |
| 22 | #10 | 60 |
| 23 | #10 | 60 |
| 24 | #10 | 70 |
| 25 | #8 | 70 |
| 26 | #8 | 70 |
| 27 | #8 | 80 |
| 28 | #8 | 80 |
| 29 | #8 | 80 |
| 30 | #8 | 90 |

**Table 9**
**Air-conditioning compressor sizing data**

## What are the requirements for heat pumps?

A heat pump is basically a central air-conditioning compressor that is used to cool the inside of a dwelling in warm weather and heat it in cold weather. This is accomplished by reversing the direction of the refrigerant through the system by means of a three-way valve in the refrigerant lines, causing the Freon® to be compressed in the inside evaporation coil (giving off heat) and to expand in the outdoor condenser coil (absorbing heat). Once the outside temperature falls below 4 °C, a heat pump is not as energy-efficient as other types of heating sources, and when outdoor temperatures drop to around 0 °C, it may not supply enough heat for a dwelling. When this happens, the heat pump is locked off by the control circuit, and another heating means (oil, gas, or electricity) in the furnace takes over. When the outside temperature warms up, the reverse occurs.

Heat pumps are installed and connected in the same way as central air-conditioning units and have the same wiring and circuit requirements (see "What are the requirements for air conditioners?" on page 143).

## What are the requirements for pools, tubs, and spas?

### What kinds of equipment are commonly installed?

The following sections survey electrical installations related to permanently installed swimming pools, storable swimming pools, decorative pools and fountains, hydromassage bathtubs, and spas or hot tubs. The Code sets out specific and often demanding requirements for these installations because the combination of voltage, current, and water around the equipment (motors, heaters, and pumps) can pose a significant safety risk. The Code requirements are intended to minimize the possibility of electric shock and electrocution.

## What are the requirements for a permanently installed swimming pool?

A permanently installed swimming pool, such as an in-ground pool, is constructed in such a manner that it cannot be readily disassembled for storage.

When any electrical equipment, such as lights, receptacles, and pump motors, is installed within 3 m (10 ft) of the inside walls of the pool, the Code requires that all the associated circuits be protected by a Class A ground fault circuit interrupter (GFCI). This is to ensure that, should the insulation on energized electrical conductors and equipment break down or the continuity of the bonding and grounding system be compromised, the potentially dangerous leakage current will be quickly shut off. The GFCI trip unit must be located more than 3 m (10 ft) away from the pool water and installed in a place that will facilitate the regular testing that these devices require. An exception to the requirement for GFCI protection is made for electrical equipment that is located behind or separated from the pool area by a fence, wall, or other permanent barrier, which would prevent a person in the pool area from contacting the equipment.

The Code does not require that any receptacles be provided near a permanently installed swimming pool. Where they are installed, however, they must be at least 1.5 m (5 ft) away from the inside walls of the pool. Receptacles located between 1.5 m (5 ft) and 3 m (10 ft) from the inside walls must have Class A GFCI protection. When receptacles are installed more than 3 m (10 ft) from the inside walls of the pool, GFCI protection is not required. To determine these distances, the measurement is taken as the shortest path that the power supply cord of an appliance connected to the receptacle would follow without piercing a building floor, wall, or ceiling. See Figure 58.

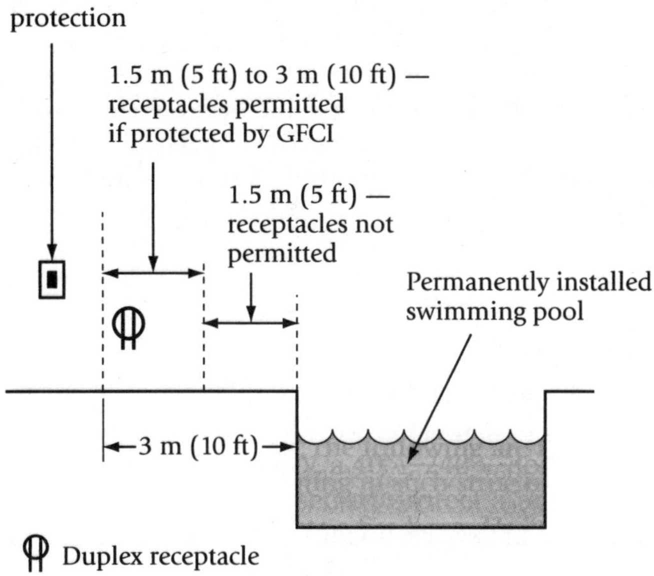


**Figure 58**
**Receptacle and GFCI protection for permanently installed swimming pools, spas, and hot tubs**

Lights are not mandatory in a pool area. When they are installed within 3 m (10 ft) of the pool surface or walls, lights and their supports or standards must, like any poolside electrical equipment, have Class A GFCI protection. See Figure 59.

Where underwater or submersible lighting fixtures are required by the owner for decorative or safety purposes, a wet-niche fixture is commonly used. The wet-niche fixture is for submersion in the pool water at a depth of no more than 600 mm (24 in), measured from the normal water level to the centre line of the wet-niche fixture lens. This 600 mm (24 in) dimension can be taken from the top of the pool deck to the centre line of the fixture lens. The wet-niche fixture is installed in an enclosure called a "forming shell," which is placed in the pool wall and connected to the deck box with a conduit. The forming shell provides a recess for supporting, holding, and protecting the wet-niche fixture. The deck box provides a place for connecting the submersible cable used to feed the underwater fixture to the branch circuit wiring. The conduit between the forming shell and the deck box gives a protected channel for the installation of the submersible cable between the deck box and the wet-niche fixture. When submersible or wet-niche fixtures are used, they must be protected by a Class A ground fault circuit interrupter.

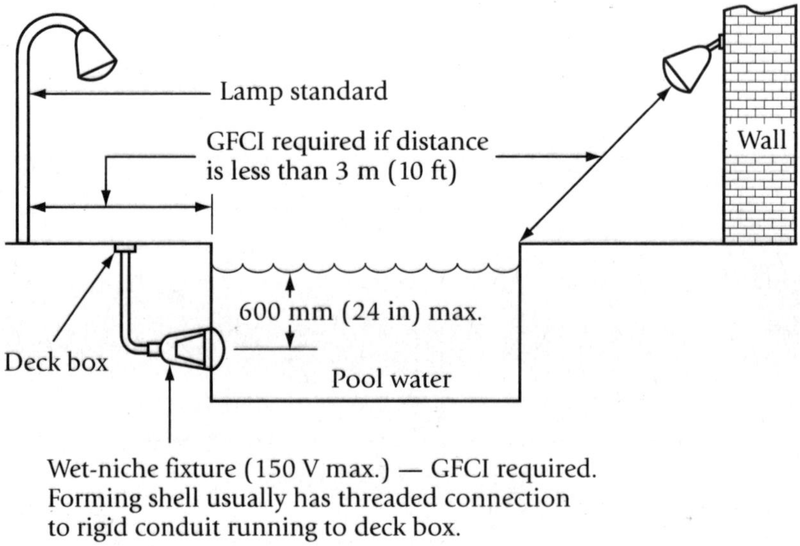


**Figure 59**
**Lighting installation for permanently installed swimming pools, spas, and hot tubs**

## What are the requirements for a storable swimming pool?

A storable swimming pool, such as an aboveground pool, is constructed in such a manner that it may be readily disassembled for storage and later reassembled.

The Code requires that no electrical equipment, such as pool pumps or lights, be located in the pool water or on the pool wall unless it has been specifically approved for this purpose. Generally, the only electrical equipment associated with a storable pool is a cord-connected electric filter pump. When the pump is within 3 m (10 ft) of the inside walls of the pool and not separated from the pool area by a permanent barrier, such as a wall or fence, the supply circuit, which may be a general-purpose branch circuit, must have GFCI protection (as described in "What are the requirements for a permanently installed swimming pool?" on page 150).

The receptacle for the pump must be installed between 1.5 m (5 ft) and 7.5 m (25 ft) from the pool. A receptacle that is located between 1.5 m (5 ft) and 3 m (10 ft) from the pool wall requires Class A GFCI protection. See Figure 60.

There are no Code requirements for mandatory lighting or any other special requirements for other kinds of electrical equipment because the pool is not a permanent installation.

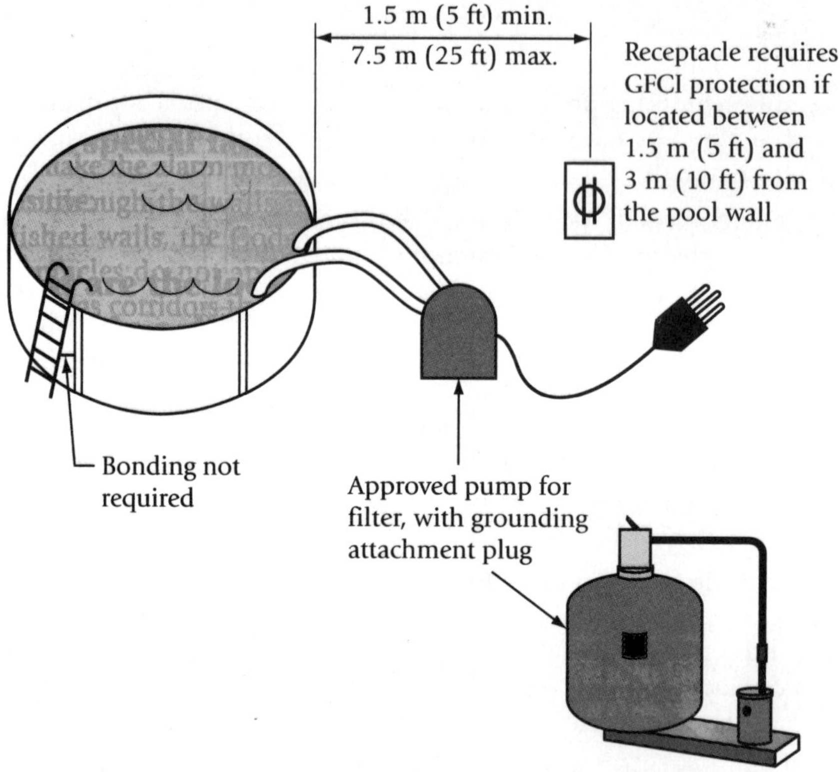


**Figure 60**
**Receptacle and pump installation for storable swimming pool**

## What are the requirements for a hydromassage tub?

The Code requires Class A GFCI protection for any electrical equipment associated with a hydromassage tub. It is the owner's or installer's responsibility to see that the tub has integral GFCI protection or is provided with GFCI protection as part of the installation (see Figure 61).

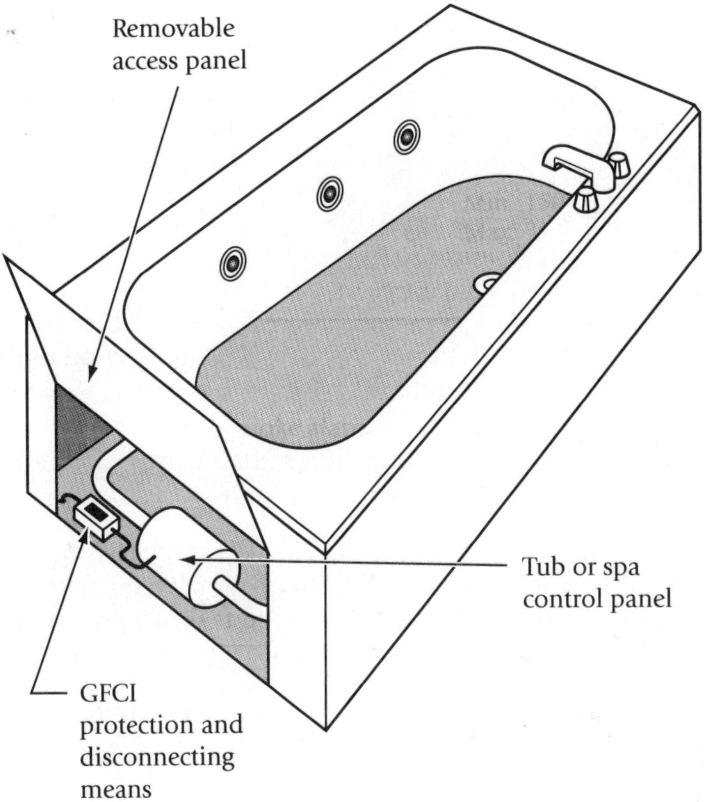


**Notes:**
**(1)** *The GFCI protection may be installed at the panel that feeds the tub.*
**(2)** *The disconnect switch must be part of the tub equipment.*

**Figure 61**
**Factory-built hydromassage bathtub, spa, or hot tub**

Methods of supplying GFCI protection include

- protecting the circuit by using a Class A GFCI circuit breaker in the panel;
- installing a "faceless" GFCI trip unit, which looks like a GFCI receptacle but has no openings for a plug; and

- installing a feed-through GFCI receptacle and connecting the hydromassage tub to the load terminals of the receptacle.

In the room or area where a hydromassage tub is installed, there is no mandatory requirement for a receptacle. If receptacles are located in the room or area, they should be, where practicable, at least 1 m (3 ft) from the tub. If this spacing is not practicable because of the room layout or other mandatory Code requirements, the distance may be reduced to 500 mm (18 in). The requirements are the same as for bathroom receptacles (see "What are the receptacle requirements?" on page 95). Accordingly, any receptacles within 3 m (10 ft) of the tub must be provided with Class A GFCI protection. The GFCI protection must not be located any closer to the tub than 1.5 m (5 ft), unless it is part of an approved factory-built tub or is located behind a barrier that will prevent an occupant of the tub from contacting it. In any case, the GFCI device must be easily available for the regular testing required by the manufacturer.

The controls for a hydromassage tub are often air switches on the tub that control the pump; these pose no shock hazard.

When electric controls are used to control the tub, the controls must be

- located behind a barrier or more than 1 m (3 ft) horizontally from the wall of the hydromassage tub (see Figure 62); or
- an integral part of an approved factory-built hydromassage tub.

Lighting and other electrical equipment or devices near a hydromassage tub must conform to the same requirements for such equipment in a bathroom. A disconnecting switch must be provided for a hydromassage tub when it is not part of a factory-built tub.

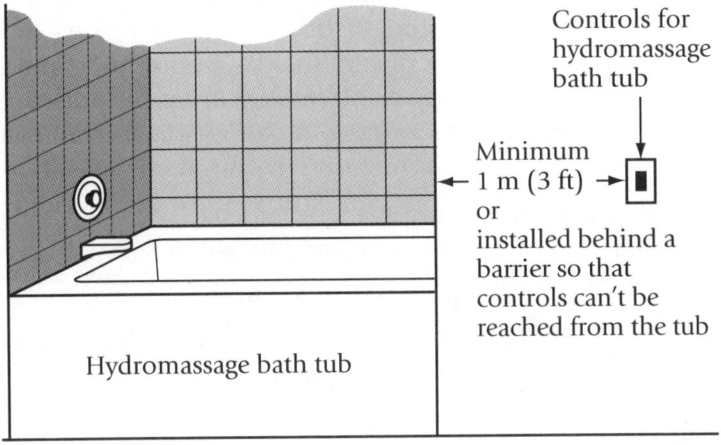


Figure 62
**Controls for hydromassage bathtubs, spas, and hot tubs (except approved factory-built units)**

## What are the requirements for a spa or hot tub?

Spas and hot tubs are similar to hydromassage tubs but are usually larger, not intended to be filled and drained with each use, and often installed outdoors. As a result, spas and hot tubs are subject to many of the Code requirements for permanently installed swimming pools. The requirements for the installation of receptacles, lights, and any other electrical equipment, except the controls, near spas and hot tubs are the same as those described in "What are the requirements for a permanently installed swimming pool?" on page 150. (See also Figure 58 on page 151 and Figure 59 on page 152.)

The controls for a spa or hot tub can be air switches that control the pump and present no shock hazard. However, when electric controls are used to control the spa or hot tub, they must be, like the controls for a hydromassage tub,

- located behind a barrier or more than 1 m (3 ft) horizontally from the wall of the spa or hot tub (see Figure 62); or
- an integral part of an approved factory-built spa or hot tub.

In any single-unit, factory-assembled spa or hot tub, GFCI protection is required for any integral electrical equipment. Many units come with GFCI protection as an integral part of the spa or hot tub control panel. The owner or installer is responsible for checking this and installing separate GFCI protection when it is not part of the factory-assembled unit (see Figure 61 on page 155).

Depending on the specific electrical requirements of the spa or hot tub, one of the following methods can be used to supply GFCI protection:

- using a Class A GFCI circuit breaker in the panel;
- using a "faceless" GFCI trip unit, which looks like a GFCI receptacle but has no openings for a plug; or
- using a feed-through GFCI receptacle and connecting the spa or hot tub to the load terminals of the receptacle.

When a spa or hot tub is field-assembled from various components, any electrical equipment installed within 3 m (10 ft) of its inside walls must also be supplied with GFCI protection. In both cases, the GFCI device must be located at least 3 m (10 ft) away from the pool water, unless it forms part of an approved factory-built spa or hot tub or is located behind a barrier that will prevent a person in the spa or hot tub from contacting it. The location must also be conveniently accessible for regular testing of the device.

The Code also requires that field-assembled spas and hot tubs be protected by leakage current collectors, devices designed to provide a path to ground for any leakage current originating from electrical devices in contact with pool water. A leakage current collector is usually a corrosion-resistant metal tube, at least five times as long as its diameter, that is bonded to ground to prevent an electrical charge on the water above ground potential (see Figure 63). The size of the bonding conductor used for the collector is based on the rating of the overcurrent device that protects the spa or hot tub (see Table 10). The conductor must be a #6 AWG copper conductor, as a minimum, if it is not protected from mechanical injury.

| Rating or setting of an overcurrent device protecting the spa or hot tub, A | Minimum copper bonding conductor size, AWG |
|---|---|
| 15 | #14 |
| 20 | #14 |
| 30 | #12 |
| 40 | #10 |
| 60 | #10 |
| 100 | #8 |
| 200 | #6 |

**Table 10**
**Minimum size of copper bonding conductor**

Leakage current collectors must be provided for all water inlets and outlets of a field-assembled spa or hot tub and installed in such a way that all water entering or leaving the spa or hot tub flows through them. These collectors are required because spas or hot tubs and their associated piping are usually made from fibreglass, plastic, or other non-metallic materials. Electric water heaters that heat the water and pumps that are used to circulate the water could put a charge on the water because of insulation breakdown or other electrical problems. Because there might be no electrically grounded metal in contact with the charged water, the charge would flow to ground through any person standing in or having contact with the spa water and touching something electrically grounded. This might not trip the circuit breaker, but depending on the level of the current, might cause injury or even death.

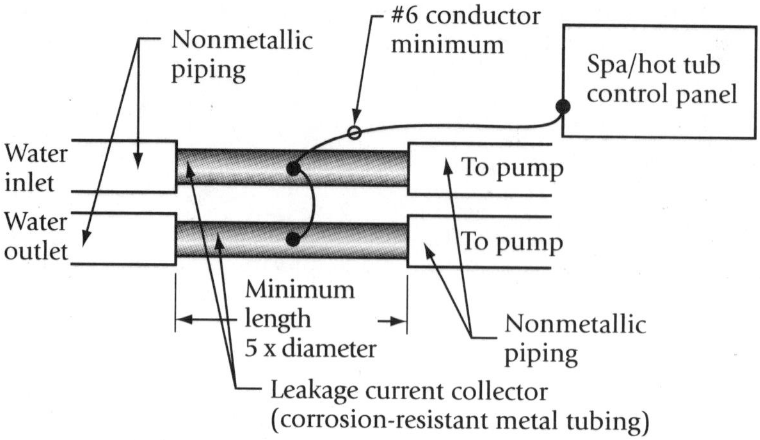


**Figure 63**
**Leakage current collectors**

Leakage current collectors need not be provided when the only electrical component in the system is a pump marked "insulated wet end pump." Because this type of electrical equipment has no metal enclosure or metallic parts in contact with the water, it will not put a charge on the water even if an electrical fault occurs.

## What are the requirements for saunas?

The room in which a sauna is installed must be built according to the specifications on the nameplate of the sauna heater. The sauna heater must be securely fastened in place and located so that the minimum clearances listed on the nameplate are maintained. The heater must not be placed below a showerhead or a water spray device, if one is provided in the sauna room.

Sauna heaters are directly connected to a supply circuit that is sized in accordance with the heater's rating (see Tables 7 and 8 on page 134). Saunas in dwellings are typically fairly small, with heaters ranging from 1 kW to 10 kW. Sauna heaters come in a wide variety of sizes to meet the different requirements posed by the size, shape, construction, and location of the sauna room. The owner or installer must know the specifications of the unit to be installed.

The Code requires that the sauna heater be controlled by a thermostat or temperature-regulating device installed according to the manufacturer's instructions. The sauna heater must be controlled by a timer, with a maximum time setting of one hour and no override feature, that disconnects all live or ungrounded conductors in the circuit supplying it. The timed switch is a safety feature that protects people in the sauna from overexposure to heat, should they become incapacitated while using the sauna. If the timed cut-off switch is not built into a sauna heater (or a factory-assembled sauna unit), it must be located on the outside wall of the room containing the sauna heater. See Figure 64.

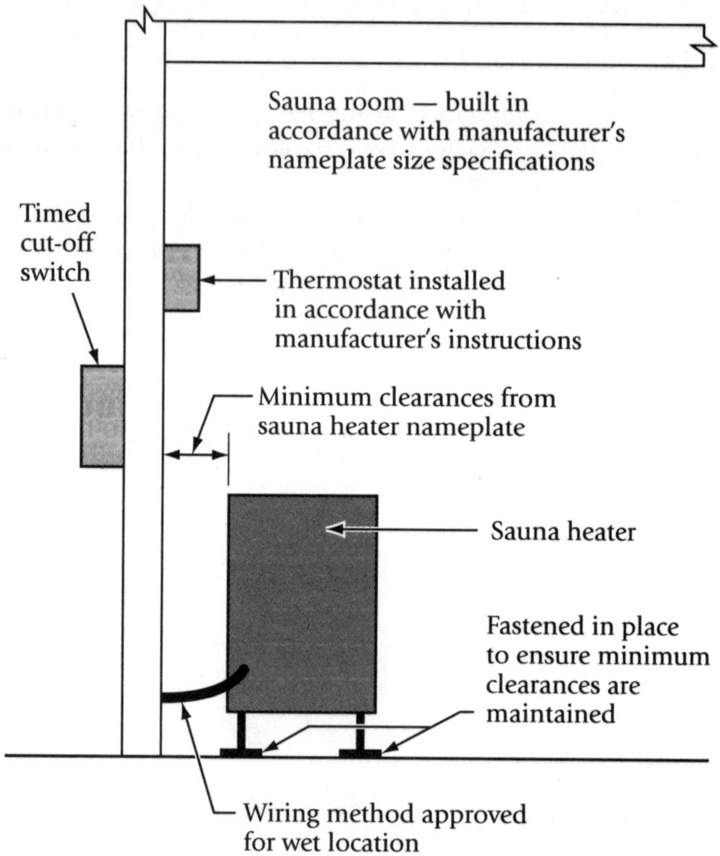


**Note:** *The sauna heater must not be installed below shower heads or water spray devices.*

**Figure 64**
**Sauna installation**

## What are the requirements for entertainment, data, and communication systems?

In dwellings, the need for outlets for communication, data, and entertainment equipment — and for periodic additions to these systems — is often overlooked. Not long ago, a few receptacles on a general-purpose circuit sufficed for one or two televisions and a stereo system or plug-in radio, and telephones ran on their own power. But now, with the proliferation of media systems (projection TVs, VCRs, DVD players, home theatre systems), the added demands of increasingly complex home computer systems (including scanners, high-end printers, networking routers, and wireless and infrared devices), as well as cordless phones, phone message machines, and fax machines, the power loads of entertainment, data, and communication systems have become considerable.

This type of equipment requires not only additional power but power of a very steady supply. It is subject to electrical disturbances that can cause data errors, poor performance, and damage to equipment or components. Such disturbances include voltage sags or swells caused by other equipment or devices on the same circuit ("normal mode disturbances"), electrical noise or dirty grounds caused by poor connections in the grounding or neutral conductors ("common mode disturbances"), and fluctuations from outside sources such as power and communication utilities.

The Code makes no specific requirements for supply circuits for this kind of equipment, but it is generally recommended that home computers and home office equipment be placed on a separate branch circuit. Because it is not always possible to anticipate where such equipment will be located, or how power needs may be affected by new equipment or components that come onto the market, it is better in new construction to make an allowance in the initial layout. It is less costly and less inconvenient to connect new appliances or equipment to an existing branch circuit than to have to install a new circuit, so it is a good idea to keep the initial loads as low as possible. Of course, there is a limit to how much load can be added, and in some cases, extensive additions or upgrades require new wiring.

Many entertainment, data, and communication systems also involve wiring that supplies outlets and connections for telephones, cable or satellite TV, computer networks, intercom systems, doorbells, and security systems. Note that the Code does not permit data, entertainment, or communication cabling to be run through the same holes in the building structure as electrical power cables. In new construction, it is usually cheaper, while the walls are open, to install the cabling for such systems and equipment, even if this equipment is not going to be installed at once. As the technology of these systems is changing rapidly, it is important to try to anticipate future requirements for cabling, for receptacles that might be needed to supply power to new equipment, and for lighting that might be needed in new work areas. Careful planning can prevent having to connect equipment with long, exposed cables and extension cords.

## What are the requirements for central vacuum systems?

In many dwellings of new construction, a central vacuum system is put in at the time of construction: the pipes and outlets are installed in the rooms or areas before the wall-finishing material is installed, and the pipes are left stubbed into the basement for future connection. The central vacuum, usually located in the basement or an attached garage, is a cord-connected appliance that must be supplied by a separate circuit, with either a general-purpose duplex 15 A, 120 V receptacle or a 20 A, 120 V T-slot receptacle (see Figure 65).

Some of these systems provide a motorized power head to supplement the central unit. In this case, receptacles should be installed adjacent to the central vacuum outlets to conveniently provide power to this equipment (see "What are the requirements for hallways and stairways in dwellings?" on page 90).

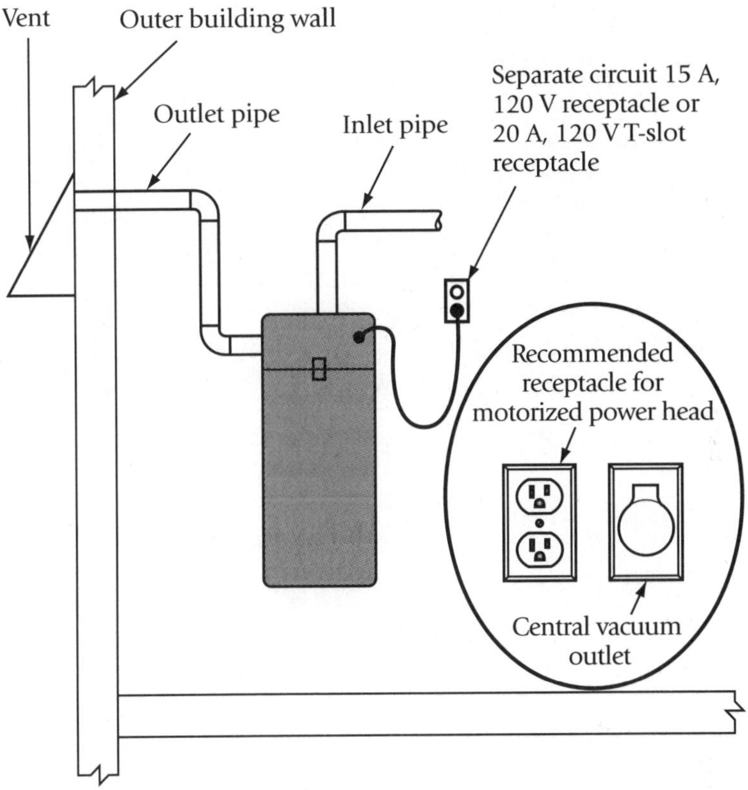


**Figure 65**
**Receptacle for central vacuum system**

## What are the requirements for motors?

In some dwellings, motors are included among the circuit loads that must be considered. These motors may power pumps (sump pumps or pressure pumps), fans, or workshop equipment (table saws, planers, or lathes), and they can be either cord-connected or direct-connected to a branch circuit.

Wherever possible — unless the motor has a very low, fractional horsepower rating — it is best to have the motor supplied by a separate circuit. If there are other loads on the same circuit, starting the motor can cause nuisance tripping of the overcurrent protection. The size of the conductors and overcurrent device for a direct-connected motor can be determined from the motor's ampere rating as shown on the nameplate (see Table 11).

If fuses are being used to provide overcurrent protection, the sizing must also take into account whether a time-delay or non-time-delay fuse is employed (see Table 11). When a motor starts, its starting surge draws six times the motor's full load current from the electrical system. A fast-acting, non-time-delay fuse in the motor circuit could take the starting surge to be a short circuit and unnecessarily open the circuit. Time-delay fuses are designed to briefly delay their tripping action, effectively distinguishing between a starting surge, which is of very short duration, and a genuine short circuit.

| Motor amperage, A | Minimum conductor size, AWG | Maximum time-delay fuse size, A | Maximum non-time-delay fuse size, A | Maximum circuit breaker size, A |
|---|---|---|---|---|
| 1–6 | #14 | 15 | 15 | 15 |
| 7 | #14 | 15 | 20 | 15 |
| 8 | #14 | 15 | 20 | 20 |
| 9 | #14 | 15 | 25 | 20 |
| 10 | #14 | 15 | 30 | 25 |
| 11 | #14 | 15 | 30 | 25 |
| 12 | #14 | 15 | 35 | 25 |
| 13 | #12 | 20 | 35 | 30 |
| 14 | #12 | 20 | 40 | 35 |
| 15 | #12 | 25 | 45 | 35 |
| 16 | #12 | 25 | 45 | 40 |
| 17 | #10 | 25 | 50 | 40 |
| 18 | #10 | 30 | 50 | 45 |
| 19 | #10 | 30 | 50 | 45 |
| 20 | #10 | 35 | 60 | 50 |

**Table 11**
**Motor sizing data**

For portable or plug-in motors, the rating of the male plug on the end of the motor's supply cord determines the minimum circuit conductor size and overcurrent protection. In this case too, if fuses are used to protect the receptacle into which the motor is plugged, time-delay fuses are recommended.

For safety reasons, a disconnect switch should be installed within sight of and not more than 9 m (30 ft) from any motor (see Figure 66). For a motor with a horsepower rating greater than 1 HP, the disconnect switch must have a horsepower rating equal to or greater than the motor's rating.

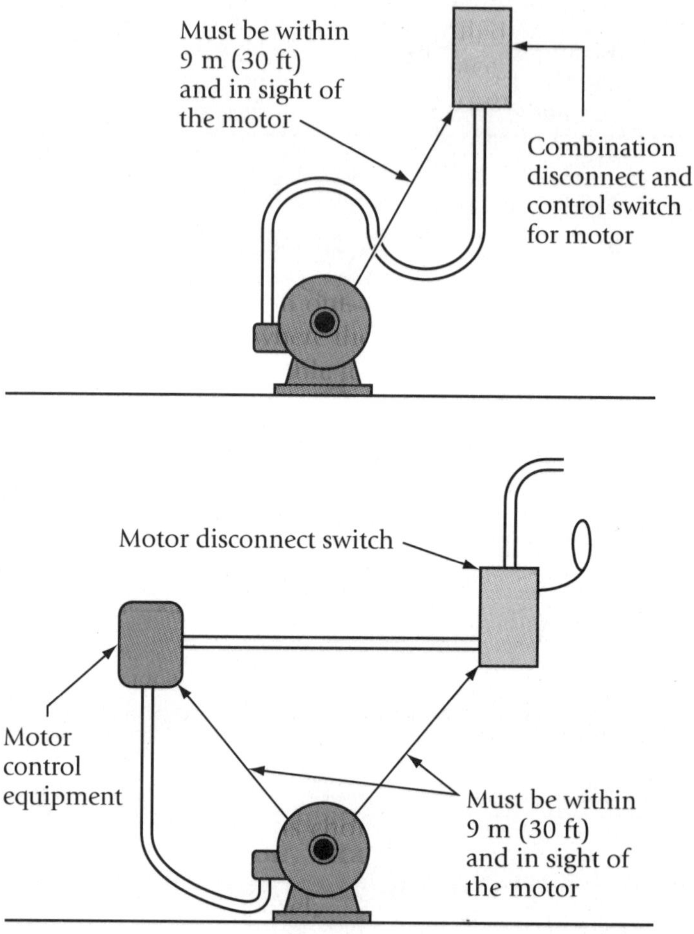


**Figure 66**
**Location of disconnect and control switches for motors**

A motor must also have excessive current (overload) protection, which may be built into the motor or supplied by the installer. In the latter case, the motor is controlled by a motor starter that incorporates an overload device to protect the motor from burnout, should the motor draw more current than it is designed to handle. A motor's overcurrent protection safeguards it only from short circuits because it must permit the very high starting surge current to flow through. (This is done by oversizing the rating of the fuse or circuit breaker on the motor circuit to 175% to 300% of the full load current, depending on the type of overcurrent device used.) To protect the motor from overheating or burnout caused by excessive current, a slow-acting overload device is used, usually installed in the controller or starter. It will not detect and respond to the starting surge but will respond to excess current over a longer period of time, de-energizing the motor before it builds up enough heat to cause damage or fire. The greater the excess current, the faster the overload device responds; the typical response time is between one and two minutes. The rating of the overload device is usually specified in the motor manufacturer's documentation. If information on this rating is not available, rating and setting of overloads in the field should be done by a qualified person; if it is not done correctly, the motor may be liable to nuisance stoppages, or it may fail to respond to excess current, which leads to overheating and possible fire hazards.

# RECEPTACLE REQUIREMENTS AT A GLANCE

| Location | Minimum requirements | Circuitry |
|---|---|---|
| Finished walls (excludes walls in kitchens, bathrooms, hallways, laundry rooms, utility rooms, and closets) | No point measured along the floor line of any usable wall space is more than 1.8 m (6 ft) from a receptacle. When a switched-duplex receptacle is used instead of a light outlet and fixture, the receptacle is considered to be one of the wall-mounted receptacles required in a finished wall or room, provided that only half of the receptacle is switched. | One of the maximum 12 outlets on a 15 A or 20 A, 120 V, two-wire branch circuit |
| Sleeping rooms and areas | Same as for finished walls | One of the maximum 12 outlets on a 15 A or 20 A, 120 V, two-wire branch circuit; must be protected by an arc fault circuit interrupter (AFCI) |
| Porch or balcony | One duplex receptacle | One of the maximum 12 outlets on a 15 A or 20 A, 120 V, two-wire circuit; ground fault circuit interrupter (GFCI) protection required if receptacle is less than 2.5 m (8 ft) from the ground |
| Dining area in kitchen | One duplex receptacle | Separate branch circuit for receptacle(s) |

*(Continued)*

| Location | Minimum requirements | Circuitry |
|---|---|---|
| Kitchen counter work surfaces | No point measured along the wall line of counter work surfaces is more than 900 mm (3 ft) from a split receptacle; measurement excludes sinks, built-in equipment, and isolated work surfaces less than 300 mm (12 in) measured along the wall line.<br>Split-duplex receptacles installed on the side of a kitchen counter designed for people with disabilities are additional receptacles and do not satisfy the requirements for receptacles on the wall. | No more than two split receptacles on a three-wire branch circuit. No other outlets should be connected to this branch circuit. Adjacent split-duplex receptacles should not be connected to the same multi-wire branch circuit.<br>A split-duplex receptacle installed on the side of a kitchen counter designed for people with disabilities may be connected to the multi-wire branch circuit supplying the two split receptacles on the wall behind the counter. |
| Kitchen refrigerator | One receptacle | Separate branch circuit that has no additional outlets other than a recessed clock outlet |
| Free-standing electric range | One 50 A receptacle mounted 130 mm (5 in) to centre above the finished floor and in the centre of the wall space left for the range. The receptacle must be a U-ground receptacle, with its slot oriented to either side. | Separate branch circuit sized by the range rating |
| Free-standing gas range | One 15 A receptacle mounted 130 mm (5 in) to centre above the finished floor and in the centre of the wall space left for the range | One of the maximum 12 outlets on a 15 A or 20 A, 120 V, two-wire circuit, or may be connected to the branch circuit for the receptacle in the dining area |

*(Continued)*

| Location | Minimum requirements | Circuitry |
|---|---|---|
| Microwave oven | In a cupboard, wall cabinet, or enclosure designated for a microwave, one receptacle is required. | Separate branch circuit |
| Hallways | No point in a hallway is more than 4.5 m (15 ft) from a duplex receptacle, as measured by the shortest path that the supply cord of an appliance connected to the receptacle would follow without passing through an opening fitted with a door. | One of the maximum 12 outlets on a 15 A or 20 A, 120 V, two-wire circuit |
| Utility room | At least one duplex receptacle | Separate branch circuit solely for the receptacles in the utility room |
| Laundry room | At least one duplex receptacle, plus one receptacle to accommodate a washing machine | Separate branch circuit solely for the receptacles in the laundry room or area |
| Bathroom (also applicable to rooms or areas that combine a sink or shower/bathing facilities with an area serving another purpose, e.g., a bedroom with an ensuite bathroom) | One duplex receptacle located within 1 m (3 ft) of the sink and, where practicable, at least 1 m, but no closer than 500 mm (20 in), from a bathtub or shower stall | One of the maximum 12 outlets on a 15 A or 20 A, 120 V, two-wire circuit; all bathroom receptacles located within 3 m (10 ft) of sinks, bathtubs, or shower stalls must be protected by a Class A GFCI |
| Combined bath and laundry room | One receptacle for washing machine located behind the machine and not more than 600 mm (2 ft) from the floor | Separate branch circuit |

*(Continued)*

| Location | Minimum requirements | Circuitry |
|---|---|---|
| Outdoors | At least one duplex receptacle not more than 2.5 m (8 ft) from ground level | Separate branch circuit protected by a Class A GFCI for all outdoor receptacles |
| Unfinished basement | One duplex receptacle | One of the maximum 12 outlets on a 15 A or 20 A, 120 V, two-wire circuit |
| Garage or carport | One duplex receptacle for each car space | Separate branch circuit, to which all receptacles, lighting fixtures, and garage door openers may be connected |
| Central vacuum system (the central vacuum duct system should be considered complete when enough ductwork has been installed to indicate the location of the central vacuum unit) | One receptacle when the duct system is installed, either a 5-15R or 5-20RA receptacle | Separate branch circuit |

# LIGHTING REQUIREMENTS AT A GLANCE

| Location | Minimum lighting requirements |
|---|---|
| Entrances | An exterior light controlled by a wall switch located inside the building is required at every entrance. |
| Kitchens, bedrooms, living rooms, utility rooms, dining rooms, bathrooms, and hallways | At least one lighting outlet with a fixture controlled by a wall switch. Bedrooms and living rooms may have a wall switch controlling a receptacle instead of a lighting outlet with a fixture. |
| Stairway lighting (other than basement stairs) | Where the stairway has four or more risers, the lighting must be controlled by a 3-way switch at the head and foot of the stairway. |
| Basement stairway lighting | Stairway lighting for basements that do not contain a finished area or room, lead to an outside entrance or built-in garage, or serve more than one dwelling may be controlled by a single-pole switch located at the head of the stairs. |
| Basements | In unfinished basements, at least one light for each 30 $m^2$ (100 $ft^2$) or fraction thereof of floor area. Once a basement room or area has been finished, the requirements for receptacles and lighting for that type of room apply. |
| Storage rooms | At least one light |
| Garages and carports | At least one light controlled by a wall switch near the doorway is required in an attached, built-in, or detached garage or carport. |

*(Continued)*

| Location | Minimum lighting requirements |
| --- | --- |
| Clothes closets | Lighting is not required in closets. If lights are installed in closets, they must be mounted on the ceiling (not above the shelf, to avoid contact with combustibles) or on the front wall of the closet above the door. Specially approved light fixtures may be mounted on the trim or sidewall of the closet doorway. Pendant or bare bulb fixtures are not permitted. |

# SMOKE ALARM REQUIREMENTS AT A GLANCE

| Location | Requirement | Circuitry |
|---|---|---|
| Floor levels | Smoke alarms must be installed on each floor level, including basements [a floor level is any level 900 mm (3 ft) or more above or below the adjacent floor level]. Each bedroom must be protected by a smoke alarm inside the bedroom or outside the bedroom within 5 m (16 ft) of the bedroom door. This measurement is made along hallways and through bedroom doorways, not through walls or openings in walls.<br>Rooms other than bedrooms may be up to 15 m (50 ft) from a smoke alarm. This measurement is made following hallways and going through doorways, not through walls or openings in walls<br>Smoke alarms must be installed on the ceiling or on a vertical wall between 150 to 300 mm (6 to 12 in) from the ceiling when the hallway or area is 1.5 m (5 ft) or less in width. | Smoke alarms may be installed in any lighting and receptacle branch circuit, except for circuits that supply kitchen counter and dining area receptacles, refrigerators, outlets in a garage, and outdoor receptacles.<br>Smoke alarms must not be connected to circuits that are protected by an arc fault circuit interrupter or a ground fault circuit interrupter or have only receptacles connected to them.<br>No switches or disconnecting means may be installed between the smoke alarm and the overcurrent device for the branch circuit.<br>Where more than one smoke alarm is required, alarms must be interconnected so that the actuation of one alarm will cause all the alarms in the dwelling to sound.<br>Battery backup is recommended. |

# CARBON MONOXIDE (CO) DETECTOR REQUIREMENTS AT A GLANCE

| Location | Requirement | Circuitry |
|---|---|---|
| Room or area that contains a solid-fuel-burning appliance | Mounted on or near the ceiling of a room or area that contains a solid-fuel-burning appliance that does not incorporate doors that substantially close off the firebox when the appliance is in operation | CO detectors may be installed in any lighting and receptacle branch circuit, except for circuits that supply kitchen counter and dining area receptacles, refrigerators, outlets in a garage, and outdoor receptacles.<br><br>CO detectors must not be connected to circuits that are protected by an arc fault circuit interrupter or a ground fault circuit interrupter or have only receptacles connected to them.<br><br>No switches or disconnecting means may be installed between the CO detector and the overcurrent device for the branch circuit.<br><br>A CO detector must have an audible alarm or be interconnected with the required smoke alarm devices in the dwelling. |

# Pool, tub, and spa requirements at a glance

| | Permanently installed swimming pool | Storable swimming pool | Hydromassage tub | Spa or hot tub |
|---|---|---|---|---|
| When is ground fault circuit interrupter protection required? | • when electrical equipment is installed within 3 m (10 ft) of the inside walls of the pool | • when the pump or pump receptacle is within 3 m (10 ft) of the inside walls of the pool | • always required on electrical equipment associated with the tub | • factory-assembled unit: always required<br>• field-assembled unit: same as for permanently installed swimming pool |
| Requirements for the installation of receptacles | • no mandatory receptacles<br>• no receptacles allowed within 1.5 m (5 ft) of the inside walls of the pool<br>• receptacles installed between 1.5 m (5 ft) and 3 m (10 ft) of the inside walls of the pool require GFCI Type A protection | • receptacle for the pump must be installed between 1.5 m (5 ft) and 7.5 m (25 ft) of the pool<br>• receptacles installed between 1.5 m (5 ft) and 3 m (10 ft) of the inside walls of the pool require GFCI Type A protection | • no mandatory receptacles<br>• no receptacles within 500 mm (20 in) of the tub, but where practicable at least 1 m (3 ft) from the tub<br>• receptacles within 3 m (10 ft) require GFCI Type A protection | • same as for permanently installed swimming pool |
| Requirements for the installation of lights | • no mandatory lights<br>• lights and supports or standards for lights installed within 3 m (10 ft) of the pool surface or walls require GFCI Type A protection | • no requirements | • no requirements | • same as for permanently installed swimming pool |

*(Continued)*

| | Permanently installed swimming pool | Storable swimming pool | Hydromassage tub | Spa or hot tub |
|---|---|---|---|---|
| Requirements for the installation of other types of electrical equipment | • GFCI Type A protection is required if the electrical equipment is located within 3 m (10 ft) of the inside walls of the pool | • no requirements | • no requirements | • same as for permanently installed swimming pool |
| Requirements for the location of ground fault circuit interrupters | • not closer than 3 m (10 ft) to the pool; location must permit testing | • same as for permanently installed swimming pool | • not closer than 1.5 m (5 ft) to the tub unless part of an approved factory-built tub or behind a barrier that will prevent the occupant from contacting it<br>• location must permit testing | • not closer than 3 m (10 ft) to the pool unless part of an approved factory-built spa or hot tub or behind a barrier that will prevent the occupant from contacting it<br>• location must permit testing |

# SIZE OF SERVICE WORKSHEETS

## Worksheet 1: Calculating the size of service for a single dwelling

| Step | Operation | Answer | Loads to be added together to determine size of service |
|---|---|---|---|
| 1 | Calculate total living area: Ground floor area + (basement floor area × 0.75) + floor area of any other living areas (including attached garage) | A Total living area = ________ $m^2$ | |
| 2 (a) | Allocate 5000 W for the first 90 $m^2$ of total living area | B 5000 W | |
| (b) | Calculate remaining living area: (A – 90 $m^2$) / 90 $m^2$ Round up answer to the next whole number. | C Remaining living area = ________ $m^2$ | |
| (c) | Calculate remaining basic loads: C × 1000 W | D Remaining basic loads = ________ W | |
| (d) | Calculate total basic loads: **B + D** | E Total basic loads = ________ W | Basic loads = **E** = ________ W |

*(Continued)*

| Step | Operation | Answer | Loads to be added together to determine size of service |
|---|---|---|---|
| 3 (a) | Allocate 10 000 W for electrical space heating (if installed) | **F** 10 000 W | |
| (b) | Calculate remaining electrical space-heating loads (if any):<br>(Rating of equipment – 10 000) × 0.75 | **G** Remaining electrical space-heating loads<br>= _________W | |
| (c) | Calculate total electrical space-heating loads:<br>**F + G** | **H** Total electrical space-heating loads<br>= _________W | |
| 4 | Determine the air-conditioning load calculated as 100% of the equipment rating | **I** Air-conditioning load<br>= _________W | |
| 5 | Add either the electrical space-heating load or the air-conditioning load, whichever is larger | | Space-heating load **H** *or* air-conditioning load **I**<br>= _________W |

*(Continued)*

| Step | Operation | Answer | Loads to be added together to determine size of service |
|---|---|---|---|
| 6 (a) | Allocate 6000 W for the first 12 000 W of the electric range rating | **J** 6000 W | |
| (b) | Calculate the remaining electric range load:<br>(Electric range rating – 12 000 W) × 0.4 | **K** Remaining electric range load<br>= ________W | |
| (c) | Add the total calculated electric range load:<br>**J + K** | **L** Total calculated electric range load<br>= _______W | Electric range load = **L**<br>= _______W |
| 7 (a) | Determine heavy equipment loads as 100% of equipment rating | **M** Electric tankless water heater load<br>= _______W<br>**N** Electric steamer load = ________W<br>**O** Electric spa load = _______W<br>**P** Electric hot tub load = _______W<br>**Q** Electric swimming pool heater load<br>= _______W | |

*(Continued)*

| Step | Operation | Answer | Loads to be added together to determine size of service |
|---|---|---|---|
| (b) | Add total heavy equipment loads:<br>**M + N + O + P + Q** | **R** Total heavy equipment loads =<br>_______W | Heavy equipment loads = **R**<br>= _______W |
| 8 (a) | Calculate the total additional loads of any equipment or appliance with a rating of 1500 W or higher | **S** Total additional loads = _______W | |
| (b) | If the dwelling has an electric range, calculate 25% of the total additional loads:<br>**S** × 0.25 | **T** 25 % of total additional loads =<br>_______W | |
| (c) | If the dwelling does not have an electric range, calculate the adjusted additional loads as 100% of each load of 1500 W or more, to a maximum of 6000 W, plus 25% of any remaining loads:<br>6000 W + [(**S** – 6000 W) × 0.25] | **U** Adjusted additional loads = _______W | |

*(Continued)*

| Step | Operation | Answer | Loads to be added together to determine size of service |
|---|---|---|---|
| (d) | Depending on whether or not the dwelling has an electric range, add 25% of the total additional loads or the adjusted additional loads | | Additional loads = **T** (dwelling with electric range)<br>= _______W<br>*or*<br>Adjusted additional loads = **U** (dwelling with no electric range)<br>= _______W |
| 9 | Estimate anticipated loads (possible renovations, additions, new equipment, etc.) as 100% of circuit and/or equipment ratings | **V** Total anticipated loads = _______W | Anticipated loads = **V**<br>= _______W |
| 10 | Total the loads and select the appropriate size of service from Table 4:<br>**E** + (**H** *or* **I**) + **L** + **R** + (**T** *or* **U**) + **V** | | Total loads<br>= _______W |

# Worksheet 2: Calculating the size of service for a multiple dwelling

| Step | Operation | Answer | Loads to be added together to determine size of service |
|---|---|---|---|
| 1 | Use "How is the size of service calculated for a single dwelling?" on page 40 and Worksheet 1 to determine the total loads for each unit, and list them in descending order. Do not include electrical heating loads or air-conditioning loads. | A Largest unit load = _______W | |
| | | B Next-largest load = _______W | |
| | | C Next-largest load = _______W | |
| | | D Next-largest load = _______W | |
| | | E Next-largest load = _______W | |
| | | F Next-largest load = _______W | |
| | | G Next-largest load = _______W | |
| | | H Next-largest load = _______W | |
| | | I Next-largest load = _______W | |
| | | J Next-largest load = _______W | |

(Continued)

| Step | Operation | Answer | Loads to be added together to determine size of service |
|---|---|---|---|
| 2 | Add the largest load | | Largest unit load<br>= **A**<br>= ________W |
| 3 | Add 65% of the loads for the next two units:<br>(**B** + **C**) × 0.65 | **K** Factored loads for units<br>= ________W | Factored loads<br>= **K**<br>= ________W |
| 4 | Add 40% of the loads for the next two units:<br>(**D** + **E**) × 0.4 | **L** Factored load for units<br>= ________W | Factored loads<br>= **L**<br>= ________W |
| 5 | Add 25% of the loads for the next five units:<br>(**F** + **G** + **H** + **I** + **J**) × 0.25 | **M** Factored load for units<br>= ________W | Factored loads<br>= **M**<br>= ________W |

*(Continued)*

| Step | Operation | Answer | Loads to be added together to determine size of service |
|---|---|---|---|
| 6 (a) | Determine the total electric space-heating loads. Allocate 10 000 W for electrical space-heating (if installed) | **N** 10 000 W | |
| (b) | Calculate remaining electrical space-heating loads (if any):<br>(Rating of equipment – 10 000 W) × 0.75 | **O** Remaining electrical space-heating loads<br>= ______W | |
| (c) | Add total electrical space-heating loads:<br>**N + O** | **P** Total electrical space-heating loads<br>= ______W | |
| 7 | Determine the air-conditioning load as 100% of the equipment rating | **Q** Air-conditioning load<br>= ______W | |
| 8 | Add either the electrical space-heating load or the air-conditioning load, whichever is larger | | Space-heating load **P** *or*<br>air-conditioning load **Q**<br>= ______W |

*(Continued)*

| Step | Operation | Answer | Loads to be added together to determine size of service |
|---|---|---|---|
| 9 | Add 94% of the total loads outside of, but shared by, the individual units:<br>Total common loads × 0.94 | **R** Common loads = _______W | Common loads<br>= **R**<br>= _______W |
| 10 | Total the loads and select the appropriate size of service from Table 4:<br>**A + K + L + M + (P *or* Q) + R** | | Total loads<br>= _______W |

# GLOSSARY

**Acceptable**
acceptable to the local authority enforcing the *Canadian Electrical Code, Part I.*

**Accessible equipment**
equipment whose accessibility is not limited by locked doors, elevation, significant obstacles, or other effective means.

**Accessible wiring methods**
wiring that is not permanently closed in by the structure or finish of a building and that is capable of being removed or repaired without disturbing a building's structure or finish.

**Accredited certification organization**
an organization that has been accredited by the Standards Council of Canada, in accordance with specific criteria, procedures, and requirements, to operate, on a continuing basis, a certification program for electrical equipment.

**Alive (live)**
electrically connected to a source of voltage difference, or electrically charged so as to have a voltage different from that of the earth. The term is often used in place of "current-carrying."

**Aluminum-sheathed cable (alu-sheath)**
cable consisting of one or more conductors and covered with a smooth or corrugated liquid- and gas-tight sheath of aluminum or aluminum alloy.

**Ampacity**
the current-carrying capacity of an electric conductor, expressed in amperes (A).

**Anode**

a rod placed in the storage tank of a water heater to protect the steel storage tank from corrosion and subsequent leaking. Because the anode rod is made of a material that corrodes before the steel tank, it should be checked every two to three years and replaced when pitted by corrosion.

**Appliance**

a portable electrical or gas-powered device or piece of equipment used for a specific task, especially for domestic tasks. Most appliances are cord-connected, not direct-connected.

**Approved electrical equipment**

- equipment certified by a certification organization accredited by the Standards Council of Canada in accordance with
  - CSA or other standards; or
  - other recognized documents, where CSA standards do not exist or are not applicable; or
- equipment in conformance with the requirements of the applicable regulatory authority.

**Arc fault circuit interrupter (AFCI)**

a device intended to provide protection from arc faults by recognizing the characteristics unique to arcing and de-energizing the circuit when an arc fault is detected. Arc faults may be caused by breaks in conductors and by damaged insulation between two insulated conductors or between an insulated conductor and ground (usually exposed grounded metal).

**Arcing**

a luminous discharge in air between two electrodes or points with different electrical charges. An example is lightning, which is an arc between clouds or between clouds and the earth. Plasma-flame temperatures associated with electrical arcing can exceed 6000 °C; hence, the use of high-temperature electrical arcing to weld metals (arc welding).

**Armoured cable**

cable consisting of one or more conductors and covered with a spiral-wound metal sheath.

**AWG**

the American (or Brown and Sharpe) Wire Gauge, a system of designating the diameter of wire by numbers.

**Bathroom**

a room that contains bathing or showering facilities and that may also contain a sink and/or toilet. (In this Guide, "bathroom" is used to refer to rooms containing any combination of bathing/showering facilities, sinks, and toilets, including those rooms commonly known as "washrooms" and "water closets.")

**Bell** or **bell-end fitting**

a flared fitting used on the end of conduit where underground direct burial conductors enter the conduit. The flared fitting gives the conduit a large, smoothly rounded opening to prevent damage to the conductors entering the conduit.

**Bonding**

providing a low-impedance path by permanently joining all non-current-carrying metal parts to ensure electrical continuity and the capacity to conduct safely any current likely to be imposed on it. Bonding connects all non-current-carrying metal parts (exposed metal that can be touched) of electrical equipment, appliances, and devices together so that there is no voltage difference between the metal parts.

**Branch circuit**

the part of the wiring installation between the overcurrent device protecting the circuit in the panel and the outlet or outlets.

**Bushing**

a sleeve providing electrical insulation and a smoothly rounded surface to prevent damage where conductors enter or exit a raceway or cable fitting.

**BX®**

a trade name for armoured cable.

***Canadian Electrical Code, Part I***

a voluntary code, published by the Canadian Standards Association, for adoption and enforcement by regulatory authorities. The Code is intended to establish safety standards for the installation and maintenance of electrical equipment.

**Certification of electrical equipment**

the procedure by which a facility, product, service, or system becomes certified in conformity with specific standards or other recognized documents.

**Circuit breaker**

a protective device that senses current flow above its allowed rating and trips or de-energizes the circuit by electro-mechanical means. Unlike fuses, when circuit breakers trip, there are no parts that need replacing. To reset the circuit breaker, wait a few minutes to allow the trip mechanism to cool and then reset the breaker. Circuit breakers can be reset by untrained people, but if the breaker repeatedly trips, it means that there is a fault in the circuit, requiring repair by a qualified person.

**Class 1 circuit**

a power circuit supplied by a power source that limits the circuit voltage to a maximum of 30 V and limits the power output to a maximum of 1000 W.

**Class 2 circuit**

a power circuit supplied by a power source that allows a higher circuit voltage than a Class 1 circuit (up to a maximum of 150 V), but restricts the current flow in the circuit to a value not exceeding 100 divided by the circuit voltage.

**Communication system**

an electrical system for receiving and/or transmitting voice, sound, or data, such as a telephone or telegraph, data communication, intercommunication, or paging systems, wired music systems, and other systems of a similar nature.

**Concealed**

made inaccessible by the structure or finish of a building.

**Conductor**

a wire, cable, or other form of metal installed for the purpose of conveying electric current from one piece of electrical equipment to another or to a ground.

**Bonding conductor**

a conductor that connects non-current-carrying parts of electrical equipment, raceways, or enclosures to the service equipment or system grounding conductor.

**Grounding conductor**

a conductor that connects the service equipment or system to the grounding electrode.

**Identified conductor**

a conductor with a white covering or, on certain flexible cords, a raised longitudinal ridge or ridges on the surface of the extruded covering to indicate that the conductor is either a grounded conductor or a grounded neutral.

**Neutral conductor**

a conductor that is always bonded to ground and intended to carry the unbalanced load between the line conductors. A grounded conductor is not always a neutral conductor.

**Conduit**

a channel of circular cross-section through which conductors may be drawn or pulled for installation or removal.

**Connector**

**Box connector**

a device for securing a cable, using its sheath or armour, where the cable enters an enclosure such as an outlet box. It may be called a 300 V, 1010, 1030, or L16 connector.

**Wire connector**

a device that connects two or more conductors together, or one or more conductors to a terminal point, for the purpose of connecting electrical circuits.

**Consumer's service**

the portion of the consumer's installation from the service box (or its equivalent) up to, and including, the point at which the supply authority makes connection.

**Cord-connected**

describes a device, appliance, or equipment that is connected to the electrical system by a suitable length of flexible cord or power supply cable with an attachment plug at the end. The attachment plug is inserted into a receptacle to provide electrical power to the device, appliance, or equipment.

**Current-carrying**

describes a circuit component made of a conductive material, such as copper or aluminum, that is designed to allow the flow of electrical current (movement of electrons) under normal operating conditions. For example, the black- and white-insulated conductors in a cable are current-carrying. They take the current from the supply to the load and back when the load is on. The bare or green-insulated ground conductor is not considered a current-carrying conductor because it has current flowing in it only under fault conditions (e.g., a short circuit, insulation failure, or leakage current).

**Damp location** — see **Location types**.

**Dead**

describes current-carrying parts of electrical equipment that are free from any electrical connection to a source of voltage, free from any electrical charge, and not having a voltage different from that of earth.

**Deck box**

a box used in the installation of a wet-niche fixture in a swimming pool, spa, or hot tub. It is placed face up on the deck of the pool, spa, or hot tub to provide a convenient means of connecting and disconnecting the cable that runs down to the wet-niche fixture below the surface of the water.

**Device box**

a box that holds a device (e.g., a switch, receptacle, or cover) by means of 6/32 inch machine screws and that is sized and identified as follows:

- 3 × 2 × 1½ in: #1100
- 3 × 2 × 2 in: #1102

- 3 × 2 × 2½ in: #1104
- 3 × 2 × 3 in: #1004

**Direct-connected**
describes equipment that is specifically designed to be used in a fixed position, has no attachment plug, and is installed to receive current in such a way that it cannot be disconnected from its source of electricity without the use of tools.

**Disconnecting means**
a device, group of devices, or other means whereby the conductors of a circuit can be disconnected from their supply source.

**Double-pole**
describes a device that is connected to two lines or energized conductors. A double-pole switch has two line connections and two load connections.

**Drip loop**
a loop left in a consumer's service conductor to prevent moisture from running along the conductor and entering the service conduit, raceway, or cable. If moisture enters the service conduit, raceway, or cable, it can flow down into the meter base or service equipment, causing the metal to corrode or the energized service conductors to short circuit.

**Dry location** — see **Location types**.

**Dwelling unit**
one or more rooms used as a housekeeping unit, with cooking, eating, living, and sleeping facilities. A dwelling unit contains a kitchen, as well as sleeping and living areas or rooms. Examples include apartment building suites, flats, and lofts.

**Electrical contractor**
a person or business that does electrical installations and any other work to which the *Canadian Electrical Code, Part I,* applies.

**Electrical equipment**

any apparatus, appliance, device, instrument, fitting, fixture, machinery, or material used in the generation, transformation, transmission, distribution, supply, or utilization of electric power or energy.

**Electrical installation**

the installation of any wiring from the point or points where electric power or energy is delivered by the supply authority to the point or points where such power or energy can be used to operate electrical equipment.

**Electrical metallic tubing (EMT)**

a metal raceway for conductors having a circular cross-section and a wall thinner than that of rigid metal conduit.

**Electrical nonmetallic tubing (ENT)**

a flexible, corrugated, nonmetallic raceway for conductors having a circular cross-section. Though sized like other circular cross-section raceways, it is intended to be used with specific types of connectors and couplings to join lengths and terminate runs of tubing.

**Enclosure types**

designations, based on environments, for boxes used for electrical devices, outlets, and connections:

- **Type 1:** for use indoors in ordinary locations;
- **Type 2:** for use indoors where the enclosure can be subject to drops of falling liquid due to condensation or other causes;
- **Type 3:** for use outdoors; and
- **Type 4:** for use where the enclosure can be subject to direct streams of water.

**Feeder circuit**

a circuit that runs between overcurrent devices.

**Fixed wiring**

the permanent wiring that is fastened to the structure of a building. It does not include appliance cords, extension cords, and flexible cords.

**Fixture**

a permanently connected unit (or group of units assembled to form a complete unit) that can be removed or replaced without damaging any part of the building structure. Examples include light fixtures, electric baseboard heaters, sauna heaters, garage door openers, etc.

**Flexible metal conduit**

a metal conduit that can easily be bent without the use of tools.

**Forming shell**

the enclosure in a pool, spa, or hot tub wall that supports and holds the wet-niche fixture in place. The forming shell is connected to the deck box with a conduit, to provide a channel for inserting or removing the underwater cable that is attached to the wet-niche fixture.

**Fuse**

a protective device that senses current flow above its rating and trips or de-energizes the circuit by melting low-temperature metal. Fuses are one-time devices. Once the low-temperature metal melts and trips the circuit, the fuse must be replaced. Spare fuses of the correct rating should always be kept in an easily accessible place near the fusible equipment. Proper procedure should be followed when replacing tripped fuses.

**General-purpose circuit**

a two-wire circuit that contains a maximum of 12 outlets (such as lighting, smoke alarms, carbon monoxide detectors), receptacles, or a combination of outlets and receptacles.

**Ground**

a connection to the earth obtained by a grounding electrode.

**Grounded**

connected to the general mass of the earth through a grounding path that has a sufficiently low impedance (resistance to current flow) and a sufficiently high ampacity (capacity for current flow) so that at all times, including under the most severe conditions liable to arise in practice, no current in the grounding conductor can cause a harmful voltage to exist between

- the grounding conductors and neighbouring exposed conducting surfaces that are in good contact with the earth; or
- the grounding conductors and neighbouring surfaces of the earth itself.

**Ground fault**

a fault that allows current to flow between an energized or line conductor and ground or grounded exposed metal. Ground faults are caused by insulation breakdown, corroded or dirty connections, moisture, or vibration. A ground fault is sometimes called a "short circuit" or "leakage current."

**Ground fault circuit interrupter (GFCI)**

a protective device that supplements the overcurrent protection of a branch circuit. GFCIs monitor leakage current (the current flowing out versus the current returning) in a circuit. When the leakage current exceeds a predetermined value, the protective device de-energizes the circuit. Unlike Class A GFCIs, the amount of leakage current required to trip the protective device is adjustable.

**Class A type GFCI**

a ground fault circuit interrupter that de-energizes the circuit when the leakage current is between 4 and 6 mA. This type of GFCI is required by the Code for most residential applications.

**Grounding**

providing a permanent and continuous conductive path to the earth. This path has sufficient ampacity to carry any fault current liable to be imposed on it and sufficiently low impedance to limit the shock hazard and to facilitate the operation of the protective devices in the circuit.

**Grounding electrode**

a buried, metal, water-piping system, or a metal object or device buried in or driven into the ground, to which a grounding conductor is electrically and mechanically connected.

**Hot tub** — see **Spa or hot tub**.

**Hydromassage bathtub**
a permanently installed bathtub having an integral or remote water pump or air blower and a fill and drain water system. Therapeutic pools are considered hydromassage bathtubs.

**Identified terminals of electrical equipment**
terminals that are silver coloured or otherwise marked, and intended for connections to grounded or neutral conductors.

**Inspection department**
an organization legally authorized to enforce the *Canadian Electrical Code, Part I,* and having jurisdiction over a specified geographical area.

**Inspector**
any person appointed by an inspection department to enforce the *Canadian Electrical Code, Part I.*

**Insulated**
separated from conducting surfaces by a material or an air space that offers a sufficiently high degree of resistance to prevent leakage of current.

**LB fitting**
a fitting used to allow conduit running along the outer surface of a wall to go through the wall with a minimum disturbance to the wall. The fitting gets its name from its "L" shape and the way the conduit leaves the access opening. In the case of an LB fitting, the letter "B" means that the conduit leaves the opening through the back.

**Leakage current**
the flow of amperes that does not return to the power source. Leakage current is determined by measuring the current flowing out from the source in a line or energized conductor to a load and subtracting the current flowing back in the neutral (white) conductor.

**Line-to-line voltage**
the voltage measured between two different line conductors. In dwellings, the typical line-to-line voltage is 240 V.

**Line-to-neutral voltage**

the voltage measured between any line conductor and the neutral (white) conductor. In dwellings, the typical line-to-neutral voltage is 120 V. If the continuity of the neutral conductor breaks down (for example, as a result of an open or high-resistance connection), the line-to-neutral voltage could fluctuate. Because many of the loads in a dwelling are line-to-neutral loads (120 V), voltage fluctuations can affect the life expectancy, durability, performance, and safety of the equipment used for these loads.

**Liquid-tight flexible conduit**

a flexible metallic or nonmetallic conduit with a liquid-tight outer jacket.

**Location types**

categories that the Code uses to distinguish varying degrees of mechanical and environmental exposure:

- **Damp location:** an exterior or interior location that is normally or periodically subject to condensation or moisture in, on, or adjacent to electrical equipment. This includes partially protected locations under canopies, marquees, roofed porches, and similar locations.
- **Dry location:** a location not normally subject to dampness. This includes locations subject to temporary dampness, as in the case of a building under construction, provided that ventilation is adequate to prevent an accumulation of moisture.
- **Ordinary location:** a dry location in which, at normal atmospheric pressure and under normal conditions of use, electrical equipment is not unduly exposed to damage from mechanical causes, excessive dust, moisture, or extreme temperatures, and in which electrical equipment is entirely free from the possibility of damage due to corrosive, flammable, or explosive atmospheres.
- **Outdoor location:** any location exposed to the weather. This does not include outdoor areas that are sheltered from the weather, such as areas under porches and soffits.
- **Wet location:** a location in which liquids may drip, splash, or flow on or against electrical equipment.

**Lug**

a device used for terminating a conductor (usually a stranded conductor or a conductor larger than #10 AWG).

**Luminaire**

a complete light fixture designed to accommodate lamps (bulbs) and to connect the lamps to circuit conductors.

**Mains**

a central distribution network for electricity.

**Make-up air fan**

a fan that brings outside air into the combustion chamber of a solid-fuel-burning or fossil-fuel-burning appliance or furnace to support the ignition and burning of solid or fossil fuels such as natural gas, propane, oil, or wood.

**Meter base**

the socket into which the supply authority installs its meter to determine the amount of electrical energy used by a dwelling. The meter base is owned by the consumer, but because the type of meter determines the type of socket/meter base that must be installed, consumers should always check with the supply authority to determine the type of meter base required.

**Mobile home**

a transportable dwelling unit constructed to be towed on its own chassis. The electrical requirements for a mobile home are the same as for a single-dwelling unit.

**Multi-wire branch circuit**

a branch circuit consisting of two or more ungrounded conductors that have a voltage difference between them, and an identified grounded conductor that has the same voltage difference between it and each ungrounded conductor. The grounded conductor is connected to the neutral conductor.

***National Building Code of Canada***

a model code, published by the National Research Council Canada, whose principal objective is to set out requirements and criteria to provide a minimum acceptable level of health and safety for occupants of buildings across Canada. Through adoption and adaptation by authorities having jurisdiction, it serves as the basis for all building regulation in Canada.

**Neutral** — see **Conductor, Neutral conductor**.

**Neutral supported cable**

a cable that has no sheath surrounding the conductors. The bare (uninsulated) neutral conductor is used to support the live insulated conductors, which are twisted around the neutral. Neutral supported cable is used only outdoors in aerial installations, usually to bring the electrical supply to a building.

**Nonmetallic sheathed cable**

a cable consisting of one or more conductors and covered with a nonmetallic outer sheath.

**Ordinary location** — see **Location types**.

**Outdoor location** — see **Location types**.

**Outlet**

a point in the wiring installation at which current is taken to supply equipment or appliances. Outlets include direct-connected lighting outlets, single and duplex receptacle outlets, smoke alarms, and carbon monoxide detectors.

**Outlet box**

a box that uses 8/32 inch machine screws for the attachment of fixtures (e.g., lights, ceiling fans, cover plates, etc.) and that is sized and designated as follows:

- 3 or 4 × ½ in deep, round: Pancake
- 4 × 4 × 1½ in deep, eight-sided: Octagonal
- 4 × 4 × 1½ in deep: Square
- 4 11/16 × 4 11/16 × 2 1/8 in deep: Square

**Overcurrent device**
a device capable of automatically opening an electric circuit under a predetermined overload or short circuit, either by fusing of metal (fuse) or by electro-mechanical means (circuit breaker).

**Overload device**
a device that offers protection from excess current, but not necessarily short-circuit protection, and that is capable of automatically opening an electric circuit either by fusing of metal (fuse) or by electro-mechanical means (circuit breaker).

**Panel (panelboard)**
an assembly of buses (solid metal conductors, usually uninsulated), connections, overcurrent devices, control equipment (with or without switches), and any other related equipment, constructed for installation as a complete unit in an enclosure.

**Permanently connected equipment**
equipment that is electrically connected to the supply by means of connectors that can be accessed, loosened, or tightened only with the aid of tools.

**Permanently installed swimming pool**
a pool constructed in such a manner that it cannot be disassembled for storage.

**Permit**
the official written permission of the inspection department, on a form provided for the purpose, that authorizes work to be started on an electrical installation.

**Phase**
describes the electrical relationship between line or energized conductors. The term "phase" is often interchanged with the term "line" (as in "line conductor" or "phase conductor").

**Portable electrical equipment**
designed not to be used in a fixed position and to receive current through a flexible cord or cable, usually with an attachment plug.

**Raceway**

a channel designed for holding wires or cables. Examples of raceways include conduit (rigid and flexible, metal and nonmetallic), electrical metallic and nonmetallic tubing, underfloor raceways, cellular floors, surface raceways, wireways, cable trays, busways, and auxiliary gutters.

**Rating, Rated**

the specific electrical characteristics (ampacity, voltage, and resistance) for which a manufacturer has designed and built a device or piece of equipment, and the characteristics used to certify the device or equipment for use in an installation.

**Readily accessible**

capable of being reached quickly for operation, repair, or inspection, without requiring a person to climb over or remove obstacles or to resort to portable ladders, chairs, etc.

**Receptacle**

a device that has one or more female contact connections (slots or holes), mounted on the same attachment bracket or yoke, to accommodate one or more male attachment plugs (blades or pins). Receptacles are often called "polarized devices," which means that they have specific terminals for specific conductors [such as live, neutral (white), ground, etc.]. The following are some examples of types of receptacles:

- **20 A T-slot:** a single or duplex receptacle that can accommodate either a 15 A or 20 A, 120 V male attachment plug. The neutral (white) female connection is in the shape of the letter "T" on its side.
- **Duplex:** two female contact connections on a mounting bracket to accommodate two male attachment plugs.
- **Single:** one female contact connection on a mounting bracket to accommodate one male attachment plug.
- **Split-duplex:** each female contact connection is connected to a different branch circuit. The line conductors to the receptacle are attached to different lines with a common neutral (white) conductor. The overcurrent protection must be tied

together so that if one line to the split-duplex receptacle is de-energized, the other line to the receptacle is automatically de-energized, avoiding a potential shock hazard.

- **Switched-duplex:** a duplex receptacle connected to one circuit, but with one of the female connections always energized and the other controlled by a switch. Switched-duplex receptacles are used in living and recreation rooms where the lighting is provided by floor or table lamps, not by a ceiling fixture.
- **Switched-split-duplex:** similar to a split-duplex receptacle, but with one of the line conductors to the receptacle always energized and the other controlled by a switch. This type of receptacle is commonly used in workshops or areas where two large wattage appliances are to be used, and one needs to be controlled by a switch that may not be part of the appliance.
- **Ungrounded (unpolarized):** used in installations until the early 1950s. These receptacles were available in single or duplex types and had no specific terminals for specific conductors.

**Residential occupancy**

the occupancy or use of a building, or a part of a building, by a person who comes and goes at will and for whom sleeping accommodation is provided. Examples include rooms in a hotel, dormitory, or retirement home, provided that these rooms do not have cooking and eating facilities. (A residential occupancy might have a mini-fridge, a microwave oven, and/or a small eating area, but it would not have a kitchen, that is, a room or area used primarily to prepare and cook food.) The term does not apply to accommodations in hospitals or correctional facilities.

**Resistance**

the property of a material that makes it resist the movement of an electrical current. Resistance creates heat.

**Rigid conduit**

an inflexible conduit of a metallic or nonmetallic material:

- **Rigid metal conduit:** an inflexible conduit of metal made to the same dimensions as standard pipe and suitable for threading and jointing with standard pipe-threading equipment.
- **Rigid PVC conduit:** an inflexible conduit made of unplasticized polyvinyl chloride that cannot be threaded in the field but is joined together by solvent-cemented, unthreaded fittings.

**Romex®**
a trade name for nonmetallic sheathed cable

**R value**
the rating of insulation with regard to its resistance to heat flow. The higher the R value, the more heat flow it can resist.

**Service box** — see **Service disconnect**.

**Service, consumer's** — see **Consumer's service**.

**Service disconnect**
a fusible switch or circuit breaker in a consumer's service installation that is used for making (starting) or breaking (stopping) and limiting the amount of electric energy that is being supplied to a building or structure directly from the supply authority's distribution system. The service disconnect has additional requirements to those of general-purpose switches or circuit breakers. The service disconnect is sometimes called the service box and can be an individual fusible switch or enclosed circuit breaker that is rated for service entrance. The service disconnect can also be a separate enclosure that is part of a combination service entrance load centre.

**Service head**
the point at which the consumer's service conductors exit the service conduit, raceway, or cable through a weatherproof fitting.

**Service locate**
the necessary information, provided by the supply authority, required to locate and install the service equipment required for the building.

**Service mast**
in overhead services, a support for the rack where the supply authority conductors terminate, installed above the dwelling roof to maintain the minimum ground clearance required by the Code and the supply authority. Service masts are made of high-tensile-strength steel tubular material.

**Service, supply** — see **Supply service**.

**Single-dwelling unit**
a dwelling unit that has access to outdoor areas at ground level and that may or may not have an attached garage or carport. A single-dwelling unit may be a detached house, a unit of a semi-detached house, a unit of row housing, or a unit of a duplex, triplex, or quadruplex house.

**Single-pole**
describes a device that is connected to one line or energized conductor. A single-pole switch has only one line connection and one load connection.

**Spa** or **hot tub**
a pool or tub designed for the immersion of persons in heated water circulated in a closed system. The system incorporates a filter, heater, and pump, with or without a motor-driven blower. A spa or hot tub is not intended to be filled and drained with each use.

**Standard outlet**
a connection point in a circuit where current is taken to supply small electrical devices, appliances, fixtures, and equipment that typically have ratings of 100 to 150 W.

**Standards Council of Canada (SCC)**
a federal Crown corporation that coordinates standardization activities in Canada and in the international standards system, including the designation of National Standards of Canada.

**Starter**
a controller, usually incorporating overload protection, designed to handle the high starting current when a motor is taken from rest to normal speed, and to handle the interrupting current characteristics when a motor is stopped.

**Storable swimming pool**

a pool constructed in such a manner that it can be readily disassembled for storage and reassembled to its original integrity.

**Stud**

a conduit running between a light fixture and the fixture coupling (hickey), or between the fixture coupling and the outlet box supporting the light.

**Supply authority**

any person, firm, corporation, company, commission, or other organization supplying electric energy.

**Supply service**

a set of conductors run by a supply authority from its mains to a consumer's service.

**Surface raceway**

a surface-mounted or pendant enclosure consisting of one or more channels to contain and protect conductors and intended to accommodate associated fittings, wiring devices, lights, and accessories.

**Switch**

an electrical device used for making (starting current flow) or breaking (stopping current flow) connections in an electric circuit. Switches must have at least the minimum voltage and current ratings of the circuit they are controlling. Switches get their name by the function they perform in an electric circuit. The following are some examples of types of switches:

- **3-way:** used in circuits where the load, usually lighting, must be controlled from two locations. A 3-way switch gets its name from the number of terminals used. A 3-way switch is sometimes called a 2-way switch because two switches are required in a circuit.
- **Flip-flop:** a switch used in an electric storage-tank water heater so that both the top and bottom heating elements in the water heater are not energized at the same time.
- **General-use:** a switch, rated in amperes and capable of interrupting its rated current at rated voltage, intended for use in general distribution and branch circuits.

- **Indicating/disconnect:** marked or designed to tell the user whether the contacts are closed (on) or open (off). When the switch is in the off (open) position, all energized conductors to the equipment are de-energized.
- **Isolating:** used to isolate a circuit or equipment from its electrical power source. Isolating switches are not rated to make (start) or break (stop) the flow of current in a circuit or to equipment.
- **Motor circuit:** a manually operated switch used to control motor loads that have a large surge of current when the motor is started. Motor circuit switches may or may not have provision for motor overload (excessive current) protection.

**Three-wire split branch circuit**
a multi-wire branch circuit with two ungrounded conductors and one grounded conductor.

**Transient voltage**
a momentary surge on a signal or power line that persists for a relatively short time and that can produce false signals or triggering impulses as well as insulation or component breakdowns and failures.

**Tripping**
the activation of an integral (internal) mechanism of a protective device designed to interrupt the electric energy in a circuit when a predetermined set of conditions have been exceeded. Tripping is intended to prevent fire and shock hazards.

**Nuisance tripping**
tripping that is not due to a fire or shock hazard, thus causing an unwarranted and inconvenient interruption in electric energy.

**Voltage**
the pressure or force that causes current flow in a conductor, expressed in volts (V):

- **Extra-low voltage:** any voltage up to and including 30 V;
- **Low voltage:** any voltage between 31 and 750 V; and
- **High voltage:** any voltage above 750 V.

**Wet location** — see **Location types**.

**Wet-niche fixture**

a light fixture intended for installation in a forming shell mounted in a pool structure, where the fixture will be completely surrounded by water.

**Wiring method**

the method used to deliver electrical energy to equipment, appliances, and devices:

- **Armoured cable (BX®) wiring methods:** a more expensive wiring method with more stringent Code requirements. Armoured cable is permitted in or on buildings or portions of buildings of either combustible or noncombustible construction. It is more expensive than nonmetallic sheathed cable, is not as user-friendly, and has more specific installation requirements. Armoured cable is commonly used in steel partitions. It should only be used by people who are trained in and qualified for using armoured cable.
- **Nonmetallic sheathed cable (Romex®) wiring methods:** the most popular and economical wiring method used in residential construction. Nonmetallic sheathed cable is designed for use in buildings of combustible construction (that is, where combustible materials are used for structural members and other building assemblies).

  In residential construction, it is increasingly common to use steel studs and framing instead of wood. The Code allows nonmetallic sheathed cables to be used as a wiring method in the steel partitions, as long as the dwelling is still defined as a building of combustible construction. However, the Code sets out very specific requirements for the use of nonmetallic sheathed cable in steel partitions. The principal concern is to prevent steel building materials, which are not grounded, from being energized by a damaged cable in contact with the steel in the partitions. Only individuals who have been trained in using nonmetallic sheathed cable in steel partitions and who have the required qualifications should use nonmetallic sheathed cable in these types of installations.

# INDEX

## C

## F



## G

# H

## I



## K



## L

## N



## O

# P



# R

## S

## T

# NOTES

# Notes

# Notes

# Notes

# NOTES

# Notes

# NOTES

# Notes

# Notes

# Notes

# Notes

CANADIAN STANDARDS
ASSOCIATION